THE NUCLEAR ARMS RACE

William Gay and Michael Pearson

The Last Quarter Century, no. 1
Edited by John H. Whaley, Jr.

American Library Association
Chicago and London 1987

Cover designed by Charles Bozett

Text designed by Ray Machura

Composed by Precision Typographers
in Times Roman and Melior on a
Quadex/Compugraphic 8400 typesetting
system

Printed on 50-pound Glatfelter, a
pH-neutral stock, and bound in
10-point Carolina cover stock
by Malloy Lithographing, Inc.
∞

Library of Congress Cataloging-in-Publication Data

Gay, William, 1949–
  The nuclear arms race.

    (Last quarter century ; no. 1)
    Includes bibliographies and index.
    1. Nuclear weapons.   2. Nuclear weapons—Bibliography.
3. Nuclear warfare.   4. Nuclear warfare—Bibliography.
5. Nuclear disarmament.   6. Nuclear disarmament—
Bibliography.   7. Arms race—History—20th century.
8. Arms race—History—20th century—Bibliography.
I. Pearson, Michael, 1945–   . II. Title.   III. Series.
U264.G39   1987   355.8'25119   86-32087
ISBN 0-8389-0467-X (pbk.)

Contents

Figures

Tables

The Last Quarter Century

For the past several decades, Americans have been debating policy on such significant issues as crime, poverty, education, the economy, agriculture, foreign aid, national defense, etc. Literally thousands of publications have dealt with these topics. Students, beginning researchers, and librarians are not necessarily helped by this information. The nuclear arms race, of which this volume is the subject, is no exception. To be useful, information must be made available as knowledge placed within a framework that permits understanding of our current social problems. Without this structure, researchers will know neither the sweep of possibilities for inquiry nor the titles most appropriate for it. As all reference librarians know, the most difficult and important step in providing assistance is getting a clear statement of need from the patron. Only then can the librarian successfully define the problem and suggest sources.

The titles in this series, The Last Quarter Century, meet the requirements of students and other researchers by providing three things: the framework, the bibliography of recommended sources, and the knowledge of how to conduct further research on the topic. *The Nuclear Arms Race* provides a succinct and clearly written overview of the topic that includes discussion of the seminal ideas and works that have shaped our thinking. Each chapter includes an annotated bibliography that emphasizes the most important works in the past decade as well as the standard titles, supplemented by a list of additional readings that the authors have found to be useful. It concludes with a guide to conducting additional research on the topic discussed.

The design of subsequent titles in the series will vary according to the subject. Regardless, the books in The Last Quarter Century series will help a reference librarian get a researcher started by providing the overview, by suggesting the more useful titles on the topic and by guiding the student through the standard reference sources and tools so that the topic can be researched further.

John H. Whaley, Jr.

Preface

During the past few decades, a number of issues have come to be seen as global social problems: environmental pollution, scarcity of non-renewable resources, economic development, and human rights. In many cases these issues have presented significant dilemmas: economic growth vs. conservation of energy resources, internal growth vs. international relations, and human rights vs. sovereignty. In each case demands for international solutions have led to an intensification of conflict between the developed and developing worlds. Concomitant with these issues but overshadowing them has been the continuing level of global militarization, especially as manifested in the nuclear arms race. Indeed, the nuclear arms race is a global social problem of unprecedented magnitude.

Six countries now have nuclear weapons; another 10-15 countries could have them by 1995. A decade ago most countries of the world agreed not to develop nuclear weapons on the provision that the United States and the Soviet Union work toward arms reduction. While there have been some agreements on nuclear arms limitation, there has been no reduction. As a result the nuclear arms race and the threat of nuclear war are most acute in terms of U.S.-Soviet relations.

Twenty years ago the United States and the Soviet Union reached the point of nuclear weapon development that each country could essentially destroy the economic and military capacities of the other. Since that time both countries have greatly increased this capacity. Today we face the development of a new generation of nuclear weapons, some deployed in Western Europe and others based in the United States and the Soviet Union.

The decisions to develop and deploy these new weapons have resulted in a series of public demonstrations in the United States and Western Europe. The June 12, 1982, demonstration in New York in support of disarmament was the largest such demonstration in U.S. history.

Hundreds of thousands have demonstrated in England, West Germany, and France. The public outcry has been directed toward not only the deployment of new nuclear weapons but also the policy of deterrence that legitimates the production and deployment of these weapons. Demands for a nuclear freeze, for a comprehensive test ban treaty, and for direct resumption of arms control talks are now discussed at the local level, in meetings of city councils, county commissions, state legislatures, seminars, and meetings of various professional societies. Groups representing physicians, lawyers, educators, and religious leaders have formed, demanding a public debate on nuclear weapons and a change in existing policy.

This book is envisioned as a primer for these debates and the need for policy change. It is a primer in two senses. First, the book covers the basic topics concerning the nuclear arms race (i.e., the book is a prim′er). Second, the book prepares the reader for research or action (i.e., the book is a pri′mer).

In terms of our first task, we seek to provide the reader with an overview of the major issues and literature on the nuclear arms race. In each chapter the text provides discussion of the issues, and the accompanying bibliography gives annotations on key texts and cites other relevant sources in the literature. The bibliography is an integral part of each chapter. The manner in which the text and the bibliographic references are arranged varies according to the nature of the topic discussed. When the topic is fairly well known, the reader is directed in the annotations to supporting literature. When the topic and its accompanying literature are less well known (e.g., strategic theory), more of the sources are incorporated in the text.

In terms of our second task, by structuring our discussion around the various debates and by pointing to the relations between the key perspectives and the positions taken on specific weapon and policy proposals, we seek to facilitate greater commitment to scholarly and political involvement in what we regard as the major social problem of this century. Nuclear weapons and the prospect of nuclear war represent threats to national security and potentially the human species. Nuclear weapons have, at minimum, significantly affected the functions and consequences of war and, by extension, the relations among nations. It is our conviction that serious study, debate, and research into these problems can lessen the current sense of futility and facilitate citizen understanding of and involvement in the political decision-making process.

To accomplish our twofold task, we have divided this book into three parts—chapters on history, consequences, and alternatives. In the first part on historical background, the issue of war itself, the development of nuclear weapons, and the subsequent policies legitimating them are examined. The second part on consequences examines the probabilities of nuclear war, the effects of nuclear weapons and nuclear war, and the cost of the arms race. The third part on deterrence and the future examines the debates about deterrence and strategic theories and explores the future alternatives open to us with or without nuclear weapons. The book ends with an extensive appendix that supports and extends the discussions provided in the text. This appendix is designed as a guide for conducting research on various aspects of the nuclear arms race.

Just how each chapter develops the subjects is seen in the following outline. Chapter 1 introduces the topic of war in a nuclear age by providing an overview of selected topics on conventional war and summarizing some of the major changes in war brought about by the development of nuclear weapons. Chapter 2 discusses the changes in weapons and supporting delivery systems since 1945. Chapter 3 presents the evolution of American nuclear policy, particularly in relation to perceptions of the Soviet Union. Taken together, these chapters provide the historical background necessary to understand the arms race and the threat of nuclear war. Chapter 4 discusses the literature on the probability of nuclear war, including scenarios for its progression. Chapter 5 follows with a review of the literature on the consequences of nuclear war, distinguishing between empirical, theoretical, and speculative data. Chapter 6 discusses the patterns and consequences of global militarization and the nuclear arms race. Chapter 7 focuses on the evolution of strategic and deterrence theories, including religious and philosophical assessments. Chapter 8 discusses alternative security futures, with and without nuclear weapons.

Acknowledgments

The authors wish to thank several people for their support and assistance in the writing of this book. First, for their continued and much appreciated support, we thank our families: Carol and Heather, and Sharon, Chad, and Jonathan. Second, we wish to thank Vicki Griffith for her patient and efficient assistance in our work. Finally special thanks go to John Whaley for his continued support and editorial comments.

Part 1

HISTORICAL BACKGROUND

CHAPTER 1

War in the Nuclear Age: Continuity and Discontinuity

War has existed, in one form or another, for thousands of years. In this century, over 50 million people have died in the two World Wars: tens of millions more have died in regional wars, civil wars, and other conflicts. War in the modern age is so destructive that many writers and politicians advocate its prohibition on the grounds that the costs far outweigh any political gains. Yet, in the period since World War II, we have seen increasing global expenditures for military activity, a continuation of conventional wars, and the proliferation of nuclear arms. Consider the following:

> Approximately 100 million people worldwide (soldiers and support personnel) are engaged in military activity.
> Military spending consumes over $600 billion annually.
> At least $50 billion are expended annually for military research and development; another $35 billion are spent in international weapons trade.
> In the past 20 years, an estimated 10 million people died in conventional wars; civilians accounted for over half of the deaths.
> Annual military expenditures average $19,300 per soldier; annual expenditures for public education average $380 per schoolage child.
> Two countries, the United States and the Union of Soviet Socialist Republics, together constituting approximately 10 percent of the world's population, expend 50 percent of the world's military budget. These two countries possess 95 percent of the world's nuclear weapons.
> The global stockpile of nuclear weapons represents about 5,300 times the total explosive force of the munitions used in World War II, a war in which at least 40 million died.[1]

Concern over these developments may be seen in recent rallies and demonstrations in the United States and Europe, in the reports issued by the United Nations, and in the public speeches by leaders of North Atlan-

---

1. Ruth Leger Sivard, *World Military and Social Expenditures 1982* (Leesburg, Va.: World Priorities, 1982), pp. 5–16.

tic Treaty Organization (NATO) and Warsaw Pact countries. Questions are being raised with increasing frequency: Are we more, or less, secure today? What are the short-run and long-run consequences of the arms race? Is a large-scale nuclear exchange possible between the United States and the Soviet Union? Has policy in the United States shifted from an emphasis on deterrence to an emphasis on winning a nuclear exchange? Are there alternative security systems to those currently in operation? Does the existence of increasing numbers of sophisticated nuclear weapon delivery systems alter the role of war in international relations? While many other questions might be added here, the point is that increasing attention is being given to the existing policies concerning war, national security, and, in particular, the significance of nuclear weapons in the world.

Current discussion about war in general and the significance of nuclear weapons falls into four main categories: patterns of conventional war, nuclear weapons, war as a component of geopolitics or international relations, and alternative forms of conflict resolution or peacemaking. The literature in the first category tends to emphasize the concepts of strategy and tactics, providing analyses of specific wars or battles, or discussions of the presumed consequences of the development of new technologies or techniques of fighting. Writings in the field of military science dominate this category. Discussion of nuclear weapons and strategy, encouraged by the recent declassification of previously secret planning documents and a popular (that is, oriented toward the non-professional) literature that questions current and projected policies or weapon developments, constitutes a second category. This category includes both the writings in the field of strategic studies, and also the more popular writings reflecting serious concern with the potential consequences of a nuclear exchange.

A third category, treating the topic of war from the perspective of politics, often overlaps the first two categories. Here war is treated as only one of the dimensions characterizing political relations among countries or groups of countries bound together by treaty or common interest. Writings in the field of international relations and foreign affairs dominate this group. And in the last category, one finds increasing attention given to works that question both the function and inevitability of war itself. Included here are writings that critique the modern structure of war as well as those that advocate nonviolent means of conflict resolution.

This book examines the issues relevant to an understanding of nuclear weapons and their significance on the threat of nuclear war. We draw on important works in each of the four categories delineated above, as it is necessary to understand that some view nuclear weapons as another weapon to be incorporated into traditional military strategy, while others argue that such weapons change the role of war in international relations. For example, many writers assume that dropping atomic bombs on Hiroshima and Nagasaki ushered in the atomic age but was not an act of nuclear war. These writers consider nuclear war as a topic that has no precedent. Others argue that we have used nuclear weapons, both in Japan and in threats over the past thirty or more years. Writers taking such a perspective are more likely to emphasize the degree to which nuclear weapons are part of ongoing military and foreign policy.

Whichever perspective one takes, the fundamental issue remains: *What is the significance of nuclear weapons?* We approach the issue in the remainder of this book through three interrelated questions: (1) How did our current dependence on nuclear weapons emerge, and how has it affected the institution of war? (2) What is the probability of nuclear war, and what would be its consequences? (3) What are the major points of view concerning nuclear weapons, and do they provide alternatives to the current dependence on nuclear weapons as a means of ensuring national security?

Before examining these questions, we consider in this chapter the relationship between conventional and nuclear war. We turn first to the varying attempts to define war. Next, we discuss the traditional perspectives on war. This is followed by a brief description of the quantitative and qualitative patterns of recent wars. Such material provides a basis for understanding some of the modern patterns of war. The initial significance of nuclear weapons is then suggested through five distinctions regarding military strategy, rules of war, conventional military forces, and consequences. It is suggested that the changes wrought by nuclear weapons have "nuclearized" conventional war preparation. We close with an outline of the major views toward nuclear weapons and the role they play in the modern world.

Conventional War: Definitions, Causes, Patterns

Definitions

As a recent literature review on war indicates, there are a bewildering number of war terms, such as limited and total war, conventional and nuclear war, preventive and pre-emptive war, wars of liberation, strategic war, guerilla war, civil war, and declared and undeclared war.[2] There are legal definitions of war, as well as definitions from the perspectives of political philosophy and the social sciences. Some definitions focus on war as a condition of those contending by force or on the

---

2. Hans van der Dennen, "On War: Concepts, Definitions, Research Data—A Short Literature Review and Bibliography," in UNESCO, *UNESCO Yearbook on Peace and Conflict Studies 1980* (Westport, Conn.: Greenwood, 1981), pp. 128–89.

objectives (such as imperialistic or counterinsurgency wars), while still others focus on the predominance of the weapons used (such as conventional or nuclear war). While various writers, including Aron (1966), Eagleton, and Schwartzenberger, see little difference between war and peace in the modern world, the predominant view follows that of von Clausewitz's classic dictum that war is "but a continuation of political intercourse, with a mixture of other means."[3]

For our purposes, war in the modern era (as distinct from the feuds, raids, and so on that were characteristic of preliterate communities) may be defined as armed conflict between specialized groups of fighters representing political communities.[4] This definition is generally consistent with von Clausewitz's definition, but it allows more elaboration. Four characteristics underlie our definition. First, war is a human phenomenon. Competition and other forms of aggression are common among various species, but war is distinctively human. Second, war involves armed conflict—the use of weapons to wound, capture, or kill opponents. Third, war involves intraspecies conflict. Among other animals, the infliction of violence by members of the same species is rare; violence directed toward members of different species is normally limited to killing prey. Fourth, war involves politically organized intergroup conflict. This characteristic distinguishes war from other forms of violence between individuals (such as murder) and from other forms of collective violence (such as rebellions or civil wars). Moreover, within the past few centuries, war increasingly has come to be associated with state sovereignty; that is, war occurs between representatives of nation-states. Nuclear war may be distinguished from conventional war by the use of nuclear weapons by at least one opponent.

Persons interested in researching the various approaches to the definition of war should begin with Wright. It remains the basic source, comparing philosophical, sociological, legal, and political definitions and approaches. Van der Dennen provides a useful compilation and suggests five categories of war concepts and definitions:

1. The first category is based on a conceptualization of the relationship between war and peace, including those concepts and definitions viewing war and peace as simply different means of achieving political ends. Examples cited include the works of Aron (1966), Kallen, Barbera, and Brodie.

---

3. Quote cited in van der Dennen, ibid., p. 128.
    4. This definition follows that of Keith Otterbein, "The Anthropology of War," in John J. Honigmann, ed., *The Handbook of Social and Cultural Anthropology* (Chicago: Rand McNally, 1973), pp. 923–58. See also the discussion in Marvin Harris, *Culture, People, Nature: An Introduction to General Anthropology*, 3rd ed. (New York: Harper, 1980), pp. 212–24.

2. A second group provide sociopolitical definitions of war that emphasize the lack of international mechanisms for peaceful resolution of conflict between states. Included here are the works of von Clauswitz, Sorel, and Aron (1966).
3. Approaches emphasizing quantitative definitions such as number of deaths or degree of participation constitute a third group. Examples include the works of Richardson, Singer and Small, and Levy.
4. The judicial conception of war views certain types of conflict as legal conditions. Included here are the works of Wright and Kelsen.
5. Legal definitions of war stress the component of state sovereignty and are reflected in the writings of Wright, Reves, Stone, and Lider.

While there is overlap between these categories, it remains evident that war can be approached from a multiplicity of perspectives.

Causes of War

Attempts to suggest the causes of war reflect the multiple approaches to its definition. Generally we may categorize perspectives into those viewing war as arising from human nature, from the organization of particular states, or from the relationships among states.[5] The first category includes micro approaches, arguing that war can be examined and explained through psychological or individual factors and characteristics. The latter two categories may be viewed as macro approaches, emphasizing structural and political bases of war.

The first general category, viewing war as part of human nature, includes a variety of approaches. Three variants may be mentioned: (1) religious writings approaching war from the view of the evil or sinful nature of human beings; (2) psychoanalytic writings emphasizing drives or innate tendencies; (3) anthropological writings centering on the animal basis of violence and aggression as responses to violations of space and territory. Taken together, these views stress an inherent ''dark'' side of human nature. Such micro approaches have received little in the way of empirical verification, but are often taken for granted in popular writings on the inevitability of human conflict. One can easily accept the pervasiveness of human conflict without generalizing into an explanation of war.

The second group of suggested ''causes'' of war emphasizes factors that predispose a society to war. Distinctive here is the emphasis on macro factors. Cultural variables such as value systems supportive of

5. The threefold categorization is adapted from Kenneth N. Waltz, *Man, the State, and War: A Theoretical Analysis* (New York: Columbia Univ. Pr., 1959). Waltz terms the categories *images* or *themes;* we have substituted the concept *cause.*

aggressive expansion or manifest destiny have been argued to be important facilitants of international conflict. Structural variables are also emphasized—for example, the inherent necessity for certain types of economies to expand markets or the inevitable outcome of maintaining large standing militaries and emphasizing war preparation in the name of national defense. These approaches have a higher level of explanatory power than the human nature category, but remain subject to serious criticisms, particularly in that they overemphasize single rather than multifactor explanations, in their tendency to apply only to certain types of wars (such as imperialism), and in their emphasis on correlates rather than necessary or sufficient conditions of war.

The dominant perspective on conventional war is found in the literature that emphasizes the central importance of the state in world politics. The state claims legitimate monopoly over the use of violence or force; its political institutions function, in part, to maintain security and national interests. War, however regrettable, is viewed as one of the means by which security and national interests are preserved. States are characterized as seeking power and protecting power. From this perspective, war and peace are different outcomes in the changing relations among states. Peace is maintained when a balance of power exists (bipolar or multipolar balance). War is functional as a means of conflict resolution. The common thread of these views is that in an anarchical world—one in which no international or supranational organization exists to deter individual states from engaging in war—war is a predictable, patterned means of conflict resolution.

It is instructive to note that most of the approaches to war accept its inevitability. War is seen to be more pervasive than peace. If war is indeed inevitable, then the formal underpinnings of nuclear doctrine (such as deterring war) are made quite problematic. To read more about the various explanations of war, see Niebuhr (1932), Ardrey, Morgenthau, Lenin, Wright, Choucri and North, Stoessinger, Dougherty and Pfaltzgraff, Woito, Waltz, and Knorr and Verba.

Quantitative Patterns of War

While war has traditionally been a favorite subject of historians, it has become a separate subject area within the social sciences only in this century. Such studies have emphasized both the quantitative and qualitative patterns in modern war. A brief examination of a few of these studies illustrates some of the changes in the character and scope of war. Sorokin examined major conflicts in Europe from the twelfth through the twentieth century and found general increases in the duration of war, size of fighting force, number of casualties, number of countries involved, and proportion of combatants to total population.[6] His statistical descriptions of war led to an increasing body of literature emphasizing quantitative patterns. These works normally focus on spatial dimensions such as extent of war, temporal dimensions (such as duration), and severity or intensity dimensions (such as total deaths and deaths proportional to population). Such studies, while providing useful general trend information, face a number of methodological problems in measurement: When is a country at war? Are specific battles considered war? Does internal rebellion constitute war? Given these shortcomings, it is instructive to mention a few of the findings of more recent studies. Wright studied changes in "warlikeness" (a nation's tendency to go to war), size of armed forces, length of wars, frequency of conflicts, and expansion (number of countries involved in a war) between 1450 and 1945 and concluded:

> War has during the past four centuries tended to involve a larger proportion of the belligerent state's population and resources and while less frequent, to be more intense, more extended, and more costly. . . . The trends . . . have together tended to concentrate military activity in time and space; to make it less easy to begin, to localize, and to end; to make it materially more destructive and morally less controllable.[7]

The findings of other studies generally support Wright's conclusions. Richardson estimated international rates of lethal conflict for the 1820 to 1945 period. He focused on changes in the frequency of various levels of violence, estimating that a total of 59 million violent deaths occurred during this period. Of this total, at least 36 million died in the two World Wars, accounting for the majority of all violent deaths during the period.[8] Levy studied war patterns among the "Great Powers" and found:

> With respect to historical trends in war, it has been found that interstate war involving the Great Powers has generally been diminishing over time. The frequency of war has been declining. Every dimension of the yearly amount of war has been decreasing except that the bloodiest years are getting bloodier. The wars that do occur have become considerably shorter, slightly lower in magnitude, and much greater in concentration. . . . Although Great Power wars have become less frequent, when they do occur they are more serious than in earlier times in every respect except duration.[9]

6. Pitirim A. Sorokin, *Social and Cultural Dynamics*, vol. 3 of *Fluctuations of Social Relationships, War and Revolution* (New York: American Book, 1937).

7. Quincy Wright, *A Study of War*, abridged by Louise Leonard Wright (Chicago: Univ. of Chicago Pr., 1964), pp. 62–63.

8. Lewis F. Richardson, *Statistics of Deadly Quarrels*, ed. Quincy Wright and C. C. Lienau (Pittsburgh: Boxwood, 1960), pp. 153–61.

9. Jack S. Levy, *War in the Modern Great Power System, 1495–1975* (Lexington, Ky.: Univ. Pr. of Kentucky, 1983), p. 170.

Finally, Sivard provides data on the estimated lethality and location of conflicts during the past twenty years. Her findings indicate that most conflict today occurs in the developing countries of the world, with the major powers involved primarily in "proxy" wars. The wars in the developing countries tend to be extended in duration and to involve as many civilian casualties as military personnel, and have resulted in a minimum of 10 million deaths over the time period studied.[10]

Qualitative Patterns of War

The quantitative studies, while providing insight into the statistical scope of modern war, do little to explain the impact of technological changes on war. Other scholars have analyzed qualitative aspects such as changes in weapons, military organization, and strategy. Wright provides important historical data in this area.[11] For each of four historical periods of modern conventional war, he characterizes changing patterns of military technique, first in relation to inventions having political and military consequences, and second, in relation to issues of strategy and the role of war in society. In the first period (1450–1648), the *adaptation of firearms* within infantry, artillery, and naval warfare significantly altered the lethality of war as well as the structure of military organization (toward disciplined infantries and light artillery and the use of broadside battleships). Dependence on armor and heavy cavalry declined, as did the use of mercenary armies. The increased lethality of military organization facilitated colonial expansion and conquests over weaker (and less armed) societies.

In the second period (1648–1789), war was altered primarily by the *professionalization of armies and navies*. With the emergence of standing military organizations (for example, by Louis XIV and Cromwell) strategy emphasized defense and fortification. Elaborate rules of strategy, siegecraft, and war conventions concerning the treatment of civilian and military prisoners developed. Officers were drawn from the nobility, conscripts from the unskilled or unemployed; property owners and peasants were generally excluded from service as they were needed for agricultural production. War casualties were primarily limited to professional military personnel and military installations. Mobilization of the population (through appeals to nationalism and patriotism) played a minor role in the waging of war as so few were directly involved.

Wright describes the third period (1789–1914) as involving the *capitalization of war*. Concurrent with the social forces of industrialization and colonialism, war in this period vacillated between dependence on a professional standing military and the mobilization of mass conscripts. Dependence on steam power, the armored vessel, and heavy ordnance

10. Sivard, *World Military and Social Expenditures 1982*, pp. 15–17.
11. Wright, *A Study of War*, pp. 64–70.

emerged, as did use of mines, submarines, and torpedoes, which significantly altered naval warfare. Weapons such as the rifle, machine gun, and other artillery underwent significant improvements in accuracy and range. These changes and adaptations depended upon the emerging industrial base of the major powers as well as on larger military budgets.

The fourth period (1914 and on) is characterized by the further *totalitarianization of war:* all phases of national life (political, industrial, psychological) organized for total war. World War I brought large-scale mobilization of the population and industrial alignment in support of military activities. Tanks and aerial bombardment during World War II facilitated an emphasis on offensive strategy, replacing the entrenchment so evident in World War I. Characteristics of modern military technique, exemplified in an intense form in World War II, include greater dependence on weapon technology; mechanization; mass mobilization and militarization of the civilian population; increased size of military forces; increased importance of industrial strength; and widespread destruction of life and property. Modern war, Wright argues, is characterized by total war, a breakdown in the distinction between military and civilian in military operations.

For persons interested in the evolution of qualitative dimensions of modern warfare, a number of other works should be consulted. Changes in weapons, military organization, and strategy are covered by Dupuy and Dupuy, Fuller, and Preston and Wise. Vagts discusses the altered roles of the military in society, particularly the equation of nationalism and militaristic values. Shifts in military strategy are described in Earle, Beaufre, and Howard. Specific interpretations of twentieth-century changes are found in Aron (1954), Brodie, and Weigley.

While there remains significant disagreement over defining or explaining war, there is general agreement over the trends in modern warfare: war has become more lethal, total in character, and dependent upon armaments. Between 35 and 50 million people died in the two World Wars of this century. Large military organizations stand in readiness to wage war. Conventional wars continue, while direct conflict between the world's superpowers has been avoided. Many would argue that this standoff is because of the development of nuclear weapons. Why?

The Significance of Nuclear Weapons

The development of the atomic bomb and its use in Hiroshima and Nagasaki ushered in the nuclear age. That the atomic bomb was used in a conventional war often obscures the fact that nuclear bombs have been used in wartime. Only over a period of years did consensus develop about the impact nuclear weapons would have on our ideas about war and national security. Five changes can be identified.

1. Toward the end of World War II, two qualitative differences emerged in the pattern of war: (a) the introduction of strategic or total bombing (using conventional bombs against the social and economic heart of the enemy), which ended the traditional distinction between military and civilian segments of the population; and (b) the introduction of nuclear weapons, which, while a continuation of the policy of strategic bombing, initiated a technique that coupled enormous *efficiency* and intense *radiation.*

The use of conventional weapons in the total bombing of Leipzig and Dresden at the end of World War II involved numerous air raids and thousands of bombs. On the other hand, Hiroshima and Nagasaki were each destroyed by a single plane dropping a lightweight atomic bomb. Such is the efficiency of nuclear weapons. In the years since the bombings of Hiroshima and Nagasaki, thousands of strategic bombs have been produced, each having the destructive yield of between 10 and 1000 times that dropped on Hiroshima.

The second qualitative difference between conventional and nuclear weapons lies in the latter's production of radiation: contaminated debris is sucked up into the atmosphere after a surface blast and falls back to earth as fallout. Beyond the initial phenomenon of radiation, which may be lethal, radiation causes long-range carcinogenic and mutagenic damage. Thus, the effects of a nuclear explosion can be passed on biologically to descendants of both combatants and noncombatants (including persons in countries totally uninvolved in the conflict situation).

2. Just as the development of major weapon innovations such as firearms and aerial warfare led to significant alterations in both the strategy and technique of conventional warfare, the development and use of atomic bombs during World War II altered subsequent war strategy. The strategy of nuclear deterrence is based on the argument that war is prevented by making the cost of such a war prohibitively high (that is, the consequences of a nuclear attack). Such a policy is theoretically possible only if the threat of attack or retaliation is seen as credible (that is, it is believed that the weapons will actually be used). Different strategies have emerged such as a counterforce strategy, wherein the military population and installations of the opponents are targeted, and a countervalue strategy, which targets the opponents civilian population and industry.

3. Changes in weapons and strategies concerning their use have also led to changes in the "rules of war." Traditional distinctions have been made between combatant and noncombatant and between limited lethal conflict and general destruction. With the development of counterforce and countervalue strategies, these traditional distinctions become blurred or are eliminated. While many recent wars have resulted in major, but incidental, civilian casualties, the basic aim of nuclear deter-

rence is to threaten the civilian population. Thus, civilian populations may live under continual fear of a sudden nuclear attack.

4. Conventional general wars (those involving total conflict between two or more countries), while devastating, have rarely resulted in the loss of more than 1 to 5 percent of the population. Material loss (for example, factories, industry, and communication systems) could be restored within a relatively short period of time. The lingering effects of radiation and potential loss of the civilian population sharply raise the costs of a nuclear war. For example, less than 2 million Americans died in World Wars I and II. Conservative estimates of casualties in a nuclear war range from 10 to 60 million American deaths. A slightly more indirect consequence, but of major importance, is the nuclear arms race. Countries with nuclear arms continually add to their nuclear arsenals, expending vast financial and technical resources in the process.

5. Conventional forces have become nuclearized: aircraft, submarines, and ground troops have nuclear weapons; nuclear options become part of conventional strategies. Thus, conventional war may escalate to the use by one party or exchange of nuclear weapons. Table 1 summarizes these changes. The transition stage, nuclearization of conventional war, is included to suggest the current position of the United States and the Soviet Union; each enters most conflict situations with the possibility of using nuclear weapons against either nonnuclear or nuclear powers.

Are Nuclear Weapons and Nuclear War Justified?

For centuries people have argued over the justification for war. The development of nuclear weapons and strategy has precipitated a new facet to the age-old debate on war. Justification of nuclear war (and the weapons needed to wage it) has been discussed only since 1945. Moreover, especially since about 1970 those who support nuclear weapons have faced the additional question of whether nuclear weapons are acceptable for first strikes against an opponent. The debate over justification revolves around four perspectives, which can be briefly described. The first two perspectives argue that we can and should live without nuclear weapons but differ on the role of war itself. The latter two perspectives argue that we must learn to live with nuclear weapons but differ over the role nuclear weapons should play in war fighting and national security.

The Pacifist Perspective

The pacifist tradition, drawing on the writings of Gandhi, Thoreau, Martin Luther King, and others, opposes all war and all weapons and advocates immediate endeavors at nuclear and conventional disarma-

Table 1.

Changes in War with the Introduction of Nuclear Weapons

	Conventional War (1450–1945)	Nuclearization of Conventional War	Nuclear War
Technique: weapons; military organization; ways of fighting	Professional standing armies; diversified military organizations; high civilian involvement in defense activities	Nuclear weapons integrated within existing military structure; sophisticated delivery systems (short-range and long-range)	Nuclear weapons shape conflict; short-term fighting; orientation toward use of nuclear weapons
Strategy: doctrine; rationales; presumed consequences	Conventional deterrence and military strength to prevail in combat; emphasis on readiness and flexibility	Nuclear deterrence; develop first and second strike policies; credible threat and continual updating of weapon systems emphasized	First and second strike strategy; control escalation and prevail in protracted conflict
War Conventions: legitimate war; rules of fighting; rules on targets	Legitimate if aggressed; war legitimate if necessary to maintain national interests, political obligations; combatant/noncombatant policy	Threat of war replaces the concept of legitimate war; combatant/noncombatant distinction blurred or made irrelevant; emphasis on limited war	Limit escalation
Consequences: lethality; environmental damage; socioeconomic; geopolitical	Increasing military casualties; wars between colonial powers and with colonies; WW I & II; emergence of total war	Importance of radiation and efficiency; carcinogenic and mutagenic effects; widespread environmental effects beyond battlefield; estimate 60–600 million casualties	Possible nuclear winter, summer; globalization of war

ment. Pacifists find no justification for weapons or war, and emphasize nonlethal means of conflict resolution: treaties, supranational political organizations, and policies of neutrality.

The Conventionalist Perspective

The conventionalist perspective includes those who accept some wars, and hence, the need for weapons, but who specifically reject any role for nuclear weapons. Representatives of this view advocate denuclearization but not universal disarmament. While agreeing with pacifists that nuclear weapons are unacceptable, conventionalists disagree over the role of conventional (nonnuclear) weapons and war. They advocate nuclear arms reduction but support the strengthening of conventional forces.

The Nuclear Deterrence Perspective

Both in policy and in the professional literature of the past thirty to forty years, the nuclear deterrence perspective is dominant. Supporters link nuclear weapons with a strategy that seeks to deter war by threatening a nuclear retaliation at levels that will assure the destruction of any adversary who initiates the use of nuclear weapons. Such a position is associated with the political policy known as "mutual assured destruction" (MAD). Nuclear weapons are justified only in retaliation and are targeted against the population and industry of the enemy. This perspective rejects first use of nuclear weapons and targeting of an adversary's military population and installations. The aims of stockpiling and developing new weapons systems are to ensure that a second strike will destroy the enemy. Arms control efforts are supported, but widespread disagreement exists over which types of arms should be developed and deployed.

The Nuclear Warrior Perspective

Particularly evident since the early 1970s, the nuclear warrior perspective rejects the argument that the development of nuclear weapons necessarily alters the strategy of war. Advocates of this position argue for the use of nuclear weapons in both first strike and second strike strategies against both military and civilian populations. The doctrine associated with this perspective is a countervailing strategy: the capacity to win or prevail in a war at whatever levels of nuclear weapon use are necessary. Waging war is viewed as an amoral act; nuclear weapons are seen as functional deterrents to aggression and as important tools with which to prevail should deterrence fail.

Bibliography

The following bibliography includes both sources cited in the chapter and related suggested readings. Important works seen as essential are annotated. The unannotated suggested readings are in three categories: general treatments, theories and causes of war, and quantitative and qualitative patterns of war. Few specific sources on nuclear war are included as they represent the major foci of Chapters 2–8.

General Treatments

Buchan, Alastair. *War in Modern Society: An Introduction.* New York: Harper, 1968.
    This book is a useful introduction, drawing together material from such fields as history, social science, and philosophy. The author summarizes three major approaches to the causes of war and the various attempts at its control. Among the specific topics treated are the changing patterns and na-

ture of war; changing strategy as a result of technological developments; and changes nuclear weapons bring in the conduct of war. The book concludes with a discussion of various measures for the control of war (disarmament, arms control, balance of power) and the limitations of existing international agencies in reducing the threat of war.

Waltz, Kenneth N. *Man, the State, and War: A Theoretical Analysis.* New York: Columbia Univ. Pr., 1959.

This book examines the role of classical political theory in understanding war. Drawing on the writings of Rousseau, Hobson, Locke, Kant, Marx, and Spinoza, Waltz identifies three images or themes on the reasons for war: human nature, the internal organization of particular states, and the relations among states (anarchy). He applies these approaches to current research and policy choices, particularly as they relate to the possibility of eliminating or controlling war.

Woito, Robert. *To End War: A New Approach to International Conflict.* 5th ed. New York: Pilgrim, 1982.

This book brings together the vast literature on war and peace in a useful, albeit unconventional, manner. The basic argument presented is that war is more of a problem than defense or security and that the conditions for ending war are known and plausible. In the first section, prevailing ideas about war and peace are described, particularly those pertinent to ending war. Seventeen subjects in world affairs are covered (including causes of war, strategy, actors in world politics, and arms control). The second section discusses a "peace initiative strategy," focusing on how a country such as the United States might move away from a dependence on war. The third section includes information on world affairs organizations and literature. The majority of the text consists of an annotated bibliography, easily one of the largest available on the subject. The bibliography is categorized into topical issues, classic statements, and works identified as focusing on the elimination of war.

Wright, Quincy. *A Study of War.* Abridged by Louise Leonard Wright. Chicago: Univ. of Chicago Pr., 1964.

This abridgement of the classic 1942 two-volume work faithfully summarizes what is perhaps the best introduction to the study of war. The original work represented a summary of over fifty studies conducted by research associates and faculty members at the University of Chicago during the period between World War I and II. The book combines historical and legal arguments with behavioral and quantitative data. The first section discusses the various approaches to the study of war and analyzes changing patterns of ancient and modern warfare. The second section looks at the conditions making for war (for example, the struggle for power, nationalism, and legal toleration of war). The third section looks at the prediction of war (probability and causes); the concluding section examines the control of war.

Aron, Raymond. *Peace and War: A Theory of International Relations.* New York: Praeger, 1966.

Barbera, H. *Rich Nations and Poor in Peace and War.* Lexington, Mass.: Lexington Books, 1973.

Barringer, Richard E. *War: Patterns of Conflict.* Cambridge, Mass.: MIT Pr., 1972.

Beer, Francis A. *How Much War in History: Definitions, Estimates, Extrapolations and Trends.* Beverly Hills, Calif.: Sage, 1974.

Bernard, L. L. *War and Its Causes.* New York: Henry Holt, 1944.

Bramson, Leon and George Goethals, eds. *War: Studies from Psychology, Sociology and Anthropology.* New York: Basic Books, 1968.

Brodie, Bernard. *War and Politics.* New York: Macmillan, 1973.

Clausewitz, Karl von. *On War.* 3 vols. London: Kegan Paul, 1911.

Eagleton, Clyde. *The Attempt to Define War.* International Conciliation Pamphlet, 1933.

Fried, Morton. *The Evolution of Political Society.* New York: Random, 1967.

Johnson, A. "War." In *Encyclopedia of the Social Sciences.* Vol. 15. New York: Macmillan, 1935.

Kallen, H. "Of War and Peace." *Social Research* (September 1939), p. 373.

Kelsen, Hans. *Law and Peace in International Relations.* Cambridge: Harvard Univ. Pr., 1942.

Levy, Jack S. *War in the Modern Great Power System, 1495–1975.* Lexington, Ky.: Univ. Pr. of Kentucky, 1983.

Lider, Julian. *On the Nature of War.* Westmead, Great Britain: Saxon House, 1977.

Nettleship, Martin, ed. *War: Its Causes and Correlates.* Chicago: Aldine, 1975.

Prosterman, Roy L. *Surviving to 3000: An Introduction to the Study of Lethal Conflict.* Belmont, Calif.: Duxbury, 1972.

Reves, Emery. *Anatomy of Peace.* New York: Harper, 1945.

Richardson, Lewis. *Statistics of Deadly Quarrels.* Edited by Quincy Wright and C. C. Lienau. Chicago: Quadrangle, 1960.

Russett, Bruce. *Peace, War and Numbers.* Beverly Hills, Calif.: Sage, 1972.

Schwartzenberger, G. "Peace and War in International Society." *International Social Science Bulletin* 2, no. 3 (1950): 336–47.

Singer, J. David and Melvin Small. *Patterns in International Warfare 1816–1965.* New York: Wiley, 1972.

Stone, Julius. *The Legal Controls of International Conflict.* London: Stevens, 1959.

Van der Dennen, Hans. "On War: Concepts, Definitions, Research Data—A Short Literature Review and Bibliography." In UNESCO, *UNESCO Yearbook on Peace and Conflict Studies 1980.* Westport, Conn.: Greenwood, 1981.

Walzer, Michael. *Just and Unjust Wars: A Moral Argument with Historical Illustrations.* New York: Basic Books, 1977.

Theories and Causes

Alland, Alexander. *The Human Imperative.* New York: Columbia Univ. Pr., 1972.

Ardrey, Robert. *The Territorial Imperative: A Personal Inquiry into the Origins of Property and Nations.* East Brunswick, N.J.: Bell, 1971.

Beres, Louis and Harry Targ. *Constructing Alternative World Futures: Reordering the Planet.* New York: Praeger, 1975.

Brodie, Bernard. *War and Politics*. New York: Macmillan, 1973.

Choucri, Nazli and Robert North. *Nations in Conflict: National Growth and International Violence*. San Francisco: Freeman, 1975.

Dougherty, James E. and Robert L. Pfaltzgraff, Jr. *Contending Theories of International Relations*. New York: Lippincott, 1971.

Dubos, Rene. *Man Adapting*. New Haven, Conn.: Yale Univ. Pr., 1965.

Falk, Richard. *A Global Approach to National Policy*. Cambridge: Harvard Univ. Pr., 1975.

_____ and Saul Mendlovitz, eds. *The Strategy of World Order*. 4 vols. New York: Institute of World Order, 1966.

Fornari, Franco. *The Psychoanalysis of War*. Bloomington, Ind.: Indiana Univ. Pr., 1975.

Freud, Sigmund. *Civilization and Its Discontents*. New York: Norton, 1963.

Hobson, J. A. *Imperialism: A Study*. Rev. ed. London: Allen & Unwin, 1938.

Hopkins, Terrence and Immanuel Wallerstein, eds. *Processes of the World System*. Beverly Hills, Calif.: Sage, 1980.

Kaplan, Morton. *Macropolitics: Selected Essays on the Philosophy and Science of Politics*. New York: Irvington, 1968.

Knorr, Klaus and Sidney Verba, eds. *The International System: Theoretical Essays*. Princeton, N.J.: Princeton Univ. Pr., 1961.

Koistinen, Paul. *The Military-Industrial Complex: A Historical Perspective*. New York: Praeger, 1980.

Lenin, V. I. *Imperialism: The Highest Form of Capitalism*. Moscow: Foreign Languages Publishing House, 1950. Reprint of 1917 edition.

Leone, Bruno, ed. *Nationalism, Opposing Viewpoints*. St. Paul, Minn.: Greenhaven, 1978.

Lorenz, Konrad. *On Aggression*. New York: Bantam, 1970.

Magdoff, Harry. *Imperialism: From the Colonial Age to the Present*. New York: Monthly Review Pr., 1978.

Melman, Seymour. *Pentagon Capitalism: The Political Economy of War*. New York: McGraw-Hill, 1970.

_____. *The Permanent War Economy: American Capitalism in Decline*. New York: Simon & Schuster, 1976.

Morgenthau, Hans J. *Politics among Nations: The Struggle for Power and Peace*. 5th ed. New York: Knopf, 1978.

Morris, Desmond. *The Human Zoo*. New York: Dell, 1970.

Nelson, Keith and Spencer Olin, Jr. *Why War? Ideology, Theory, History*. Berkeley, Calif.: Univ. of California Pr., 1979.

Niebuhr, Reinhold. *Moral Man and Immoral Society: A Study of Ethics and Politics*. New York: Scribner, 1932.

_____. *The Structures of Nations and Empires: A Study of the Recurring Patterns and Problems of the Political Order in Relation to the Unique Problems of the Nuclear Age*. New York: Kelly, 1959.

Rosen, Steven, ed. *Testing the Theory of the Military-Industrial Complex*. New York: Lexington, 1973.

Schell, Jonathan. *The Fate of the Earth*. New York: Knopf, 1982.

Sorel, George. *Reflections on Violence*. New York: Macmillan, 1950 (1906).

Stoessinger, John. *Why Nations Go to War*. New York: St. Martin's, 1974.

Wallerstein, Immanuel. *The Modern World System II: Mercantilism and Consolidation of the European World Economy, 1600–1750.* New York: Academic, 1980.

Waltz, Kenneth. *Theory of International Relations.* Reading, Mass.: Addison-Wesley, 1979.

Quantitative and Qualitative Patterns of War

Aron, Raymond. *The Century of Total War.* Garden City, N.Y.: Doubleday, 1954.

Beaufre, Andre. *Deterrence and Strategy.* New York: Praeger, 1966.

_____. *Introduction to Strategy.* New York: Praeger, 1965.

Brodie, Bernard. *War and Politics.* New York: Macmillan, 1973.

Dupuy, R. Ernest, and Trevor Dupuy. *The Encyclopedia of Military History from 3500 B.C. to the Present.* New York: Harper, 1977.

Earle, Edward, ed. *Makers of Modern Strategy: Military Thought from Machiavelli to Hitler.* New York: Atheneum, 1970.

Fuller, J. F. C. *A Military History of the Western World.* 3 vols. New York: Funk & Wagnalls, 1954.

Howard, Michael. *Theory and Practice of War.* London: Cassell, 1965.

Levy, Jack S. *War in the Modern Great Power System, 1495–1975.* Lexington, Ky.: Univ. Pr. of Kentucky, 1983.

Preston, Richard and Sydney Wise. *Men in Arms: A History of Warfare and Its Interrelationships with Western Society.* 4th ed. New York: Holt, 1979.

Singer, J. David and Melvin Small. *The Wages of War, 1816–1965: A Statistical Handbook.* New York: Wiley, 1972.

Sivard, Ruth. *World Military and Social Expenditures, 1983.* Leesburg, Va.: World Priorities, 1983.

Small, Melvin and J. David Singer. *Resort to Arms: International and Civil Wars, 1816–1980.* Beverly Hills, Calif.: Sage, 1982.

Sorokin, Pitirim. *Social and Cultural Dynamics.* 4 vols. New York: American Book, 1957.

Vagts, Alfred. *A History of Militarism: Civilian and Military.* New York: Meridian, 1959. Revised edition of 1937 publication.

Weigley, Russell. *The American Way of War: A History of United States Military Strategy and Policy.* Bloomington, Ind.: Indiana Univ. Pr., 1977.

CHAPTER 2

The Evolution
of Nuclear Weapons

The nuclear arms race began early in World War II and has expanded to a point where about 50,000 nuclear weapons are stockpiled globally. Current plans would increase this number by up to 50 percent before the end of the century; most of this increase would come from further stockpiling by the United States and the Soviet Union, which currently control about 95 percent of all nuclear weapons. At the same time, China, France, Great Britain, and probably India and Israel are now members of the "nuclear club";[1] another fifteen or so countries may develop nuclear weapons by the end of the century. The quantitative and qualitative arms competition between the two dominant powers reflects technological, doctrinal, and policy shifts over the past few decades. The causes and consequences of such competition have generated great concern in many quarters, resulting in increased debate over policy, weapons, and strategy. Until recently, this debate has been primarily the province of specialists in nuclear strategy, military policy, and arms control. Today, it is a topic with a broader audience and interested public.

In this chapter we review selectively the historical development of nuclear weapons. This review includes a discussion of the major events (such as the Manhattan Project and Hiroshima and Nagasaki) and changes in weapons systems (for example, Triad, MIRV), and also contains an overview of the contrasting interpretations and debates over nuclear weapons. Chapter 3 then summarizes the major policies that have emerged concerning the use of nuclear weapons. The scholarly works relevant to an understanding of the evolution of nuclear weapons

---

1. India detonated a nuclear device in the early 1970s but has denied interest in manufacturing nuclear weapons. See *Report of the Secretary-General of the United Nations, Nuclear Weapons* (Brookline, Mass.: Autumn Pr., 1980), pp. 155–59. More recently, reports have surfaced that claim Israel has secretly been producing nuclear weapons for the past twenty years (*New York Times,* Nov. 16, 1986, E-3).

are diverse and immense; they include biographies and memoirs of diplomats and scientists, internal security and military documents, congressional hearings, and academic studies. While many books and articles include short discussions of the history of the arms race, the vast majority tend to be ahistorical, repetitive, and subject to an imprecise terminology. At the same time, information on nuclear weapons is accessible and a matter of little controversy. Still, the reader with limited time and expertise desiring an understanding of the antecedent conditions to the current debate over nuclear weapons faces an imposing task when confronting the literature. The problem is how one can find and integrate the bits and pieces of relevant information distributed among a large number of highly technical yet time-specific sources.

This chapter is designed to help the reader bypass this problem. We have integrated reference to the key texts into a generally historical narrative that is organized around the major topics and breakthroughs that have occurred. Where appropriate, conceptual distinctions are provided to clarify the quantitative from the qualitative dimensions of the arms race. Moreover, while development of weapons by other countries is outside the perspective of this book, some information is provided on the weapon status of other countries constituting the nuclear club.

Early Developments: From the Manhattan Project to the Hydrogen Bomb

The splitting of the atom remains one of the great stories of modern science. Scientists throughout much of the Western world played important roles in the late 1800s and the early decades of this century: Faraday's argument that the atom had particles of electricity, the work of the Curies and Becquerel in the discovery of uranium, Bohr and Rutherford's work with radioactive particles, Chadwick's discovery of the neutron, and so on. Building on these earlier advancements, scientists in the 1930s developed the cyclotron and particle accelerator, which led to Fermi's splitting of the atom (subsequently refined by the German physicists Hahn and Strassman's discovery of atomic fission). The discovery in 1940 that uranium 235 could produce a chain reaction, followed in 1941 by the isolation of plutonium that would fission more readily than uranium 235, provided the important scientific breakthroughs necessary for the development of the atomic bomb.

Fearing that the Germans were close to developing an atomic bomb (a fear that was later found to be untrue) and aware of the significance of such a bomb, President Franklin Roosevelt in 1942 established the Manhattan Project under the leadership of General Leslie Groves and Robert Oppenheimer. This project, which culminated in the successful development of an atomic explosion, remains a topic of great interest. It was, perhaps, the first and most concentrated, large-scale scientific and military experience in weapon development: thousands of scientists and military personnel were brought together under conditions of intense secrecy to produce what has come to be seen as *the* weapon of mass destruction.

The writings on the Manhattan Project vary widely, from personal memoirs to journalistic accounts to formal reports only recently published. The official history is presented in Hewlett and Anderson, Smyth, and Brown and McDonald. Memoirs by insiders to the project are presented in Lilienthal, Compton, and Groves. A more recent recounting of the project, particularly useful in its depiction of the major personalities involved, is found in Dyson. Two of the more useful general accounts are found in Lamont and Laurence. Among the excellent audiovisual material on the subject are the Time-Life documentary and the radio series produced by the group SANE (Committee for a SANE Nuclear Policy).

The successful test of the atomic bomb in July 1945 occurred as the major Allied leaders were meeting in Potsdam to plan the invasion of Japan. While the scientists involved in the Manhattan Project disagreed as to the proper use of the bomb (whether to use it as a demonstration, or on military targets only, or on the civilian population), the political debate over the bomb centered around President Truman's belief that the dropping of the bomb would both lessen the necessity for the invasion of Japan and send a symbolic message to Stalin that the United States had unprecedented weapons and would use them: the Soviets would have to curtail expansionist tendencies or face a genuine threat from the United States. Analyses and memoirs indicate little, if any discussion over the decision to drop more than one bomb (Alperowitz, 1965; Smyth, 1945; Armine, 1959; Dyson, 1979; Stimson and Bundy, 1948). Given the destructiveness of the bomb, it is both surprising and significant that such discussion was absent, except among a few of the scientists involved in the Manhattan Project.

On August 6, 1945, the first atomic bomb used in warfare was dropped on Hiroshima. Code-named "Little Boy," the bomb had an explosive yield equivalent to 12,500 tons of TNT and killed 140,000 people. On August 9, a second bomb was dropped, this time on Nagasaki, with an explosive yield equivalent to 22,000 tons of TNT and killed 70,000 people. The debates over these uses of the bomb are presented in Alperowitz, Feis, Armine, and Giovannetti and Freed. A particularly useful reader on the subject is that edited by Fogelman. The decision to use atomic weapons against Hiroshima and Nagasaki rested on the conviction that American lives would be saved (casualties to be incurred with an invasion) and Japan would be forced to surrender. The effect of these two bombs was seen as strategically equivalent to the thousands of bombs dropped in the total bombing of German cities.

Recently, understanding of the consequences of the bombs dropped on Hiroshima and Nagasaki has been updated and extended considerably (see Chapter 5) with the release of the findings of the Committee for the Compilation of Materials on Damage Caused by the Atomic Bomb. This report synthesizes thirty years of research on the various consequences of the bombings and extends the earlier work of Oughterson and Shields, and the report of the U.S. Strategic Bombing Survey. Works by Lifton and Hersey, while fundamentally different in tone and scope, remain two of the most widely read accounts of the psychological impact of the bombings. The political significance of the bombings is discussed in Stimson and Bundy, Sherwin, and Blackett.

At the end of World War II only the United States had atomic weapons, and these few atomic bombs had to be delivered by aircraft. Over the next few years, two events were decisive in the history of weapon development. First, the United States lost its monopoly of atomic weapons in 1949 with the successful test of an atomic bomb by the Soviet Union, followed two years later by Great Britain. Second, the United States decided to produce a second and vastly more destructive weapon—the hydrogen bomb. Using a fission explosive to generate thermonuclear processes, the hydrogen or fusion bomb represented about as large a jump in destructive power from the atom bomb as did the atom bomb from conventional bombs. (See Glossary for fusion-fission distinction.) The Soviet Union successfully tested a thermonuclear bomb in 1953, the year following the American test.

The literature on the decision to build the hydrogen bomb generally focuses on the internal political and scientific debates. The possibility of a fusion bomb had been recognized early in the Manhattan Project. After World War II, many scientists, Edward Teller in particular, promoted the hydrogen bomb as necessary to maintain superiority over the Soviet Union. Others, particularly J. Robert Oppenheimer, argued against the new bomb, noting that it was not needed for military targets and would only further the policy of exterminating civilian populations. Lilienthal (1964) and Kennan (1958) claimed that a "super bomb" would make the United States overly dependent upon weapons of mass destruction. Hydrogen bomb supporters, backed by public opinion, prevailed with the argument that development was necessary in order to stay ahead of the Soviet Union. This decision is best presented in Hewlett and Duncan, and in Schilling's article "The H-Bomb Decision: How to Decide without Actually Choosing." The debate between Oppenheimer and Teller is analyzed in York (1976). Other relevant works include Davis (1968), Moss, and Lens.

Development of Strategic and Nonstrategic Weapons

By 1950, President Truman had given his approval to (1) a policy of containment toward the Soviet Union; (2) increased military spending on conventional forces; (3) production of the hydrogen bomb; and (4) an announcement that the United States was prepared to use nuclear forces to deter a Soviet attack. At this time, nuclear weapons still could be delivered only by long-range bombers (that is, the Strategic Air Command [SAC]). Two reports presented to the National Security Council provide important evidence on the rationale for further weapons systems development. The Technological Capabilities Panel of the Scientific Advisory Committee of the Office of Defense Mobilization issued the first report, *Meeting the Threat of Surprise Attack* (often referred to as the Killian Report after its chairman). The report emphasized the offensive capacity of the United States but warned that existing bombers were vulnerable to Soviet attack. It made recommendations to protect SAC bases and speed production of the intercontinental ballistic missile (ICBM), which would provide a second delivery system.

The Security Resources Panel issued the second report, entitled *Deterrence and Survival in the Nuclear Age* (often referred to as *The Gaither Report*). Also warning of vulnerability, this report recommended protection of existing bases and an early warning capability and concluded, "There will be a continuing race between offense and defense. Neither side can afford to lag or fail to match the other's efforts. There will be no end to the technical moves or counter-measures."[2] These reports reflect the early concerns over vulnerability and increases in offensive capacities. Most of the work on weapons and delivery systems in the subsequent twenty-five to thirty years reflects different responses to these concerns. Two of the more important developments involve completion of three independent systems of strategic forces (Triad) and the testing and deployment of nonstrategic weapons.

Triad

Throughout the 1950s, the United States retained nuclear superiority over the Soviet Union, but understood it could not maintain the clear monopoly of the late 1940s. Instead, planners developed diversified delivery systems—land, sea, and air—on the theory that at least one leg of this triad would be capable of retaliation in case of a surprise attack. Technological developments in ballistic missiles and jet propulsion expanded delivery systems. The Navy armed its submarines and carrier-based jets with nuclear weapons; the Air Force deployed land-based missiles while

2. Lawrence Freedman, *The Evolution of Nuclear Strategy* (New York: St. Martin's, 1981), p. 161. For a full copy of the report, see U.S. Congress, Joint Committee on Defense Production, *Deterrence and Survival in the Nuclear Age* (The Gaither Report of 1957), Joint Committee Print (Washington, D.C.: Govt. Print. Off., 1976).

continuing to rely upon long-range bombers. The development of ICBMs by both the Soviet Union and the United States dramatically increased the accuracy and potential effectiveness of the nuclear arsenals and presented significant problems to any desire for defense. Each of the air, land, and sea-based systems was able to deliver longer range nuclear weapons to its targets. Collectively those systems are referred to as Triad.

Nonstrategic Weapons

A second development was the production of nonstrategic (tactical or theater) nuclear weapons. The term *strategic* came to be applied to weapons targeted on an enemy's home-based military installations (termed strategic counterforce weapons) or civilian populations (termed strategic countervalue weapons). American deployment in 1953 of the "Honest John," the first short-range nuclear missile, introduced more flexibility in nuclear war planning. Designed as a deterrent to a Soviet invasion of Western Europe, these *tactical* or *battlefield* nuclear weapons had smaller yields and range (they could not reach the Soviet Union). A further distinction came to be made with the deployment in the late 1950s of intermediate-range Thor and Jupiter missiles in Europe. By the mid-1960s strategic (long-range) and nonstrategic (short-range and intermediate-range) weapons could be delivered by air, land, and sea.

By the early 1960s, although the Soviet Union was perceived to be ahead in certain technologies, the United States still remained dominant in the size of its arsenal: at the end of 1961, the American inventory of strategic weapons included 1,700 strategic bombers, 63 ICBMs, and 96 submarine launched ballistic missiles (SLBMs). In contrast, the Soviet Union's inventory included only 160 to 190 bombers, 20 operational ICBMs, 1200 medium-range bombers, and 600 intermediate-range ballistic missiles; the latter two systems were targeted against Western Europe.[3]

Declassification of internal policy documents of the National Security Council, Departments of Defense and State, and other government bodies has furthered explanation of the early development of strategic and nonstrategic weapons. The Etzold and Gaddis collection provides a wide-ranging series of internal documents for the researcher interested in early discussions of the atomic and hydrogen bombs. Mandelbaum and Freedman both describe the conflicts between the military services over the use of nuclear weapons and the reasons for development of nonstrategic weapons. Barnet places these conflicts within the context of NATO's emergence in the late 1940s. General treatments on the ele-

3. Union of Concerned Scientists. "A Chronology of the Nuclear Arms Race 1945 to the Present," *Briefing Manual: A Collection of Materials on Nuclear Weapons and Arms Control* (Cambridge, Mass.: Union of Concerned Scientists, 1983), p. 23.

ments of Triad are covered in Beard, Halperin (1967), and the Defense Department's annual reports on research, development, and acquisition.

Discussions of nonstrategic weapons are generally imbedded within analyses of NATO policies. Debate over their use is found in the positive assessments of Gray and of Van Cleave and Cohen. More critical views are presented by Dyer and Brenner.

Nuclear Weapons in the Modern Era

Modern weapon systems consist of various major components, including the warhead, the delivery vehicle, the command, control, and communications equipment by which the system is activated and directed.[4] There is a continuous evolution of the various elements of a weapons system. At the same time, new weapons systems are also always being developed, often taking seven to ten years from initial research to deployment. Between the United States and the Soviet Union, neither the areas nor the speed of development are synchronized. What is often referred to as the arms race includes both the evolution and refinement of existing systems and the development of new systems. In the area of nuclear technology, qualitative innovations are often followed by quantitative production. For example, once the qualitative development of atomic and then hydrogen weapons had been achieved, their quantitative stockpiling could be pursued. Moreover, as improvements occur within a country's arsenal, countermeasures may be developed by adversaries, which may in turn lead to responses to the countermeasures.

By the early 1960s both the United States and the Soviet Union were intensifying production of nuclear arsenals and diversifying their delivery systems (the triad of air, sea, and land bases). At the same time, defense planners searched for countermeasures to the buildup of strategic weapons. Two types of countermeasures can be taken against incoming warheads: passive defense (for example, civil defense programs) and active defense (for example, antiballistic missile systems).

Civil Defense

Concern for civil defense in the United States appeared in three phases.

1. Beginning with the Civil Defense Act of 1950, the initial phase focused on the Evacuation Route Program. Plans were developed to evacuate major urban areas and presumed a twelve-hour lead time (the length of time existing bombers took to reach American targets); additionally, schools and other facilities provided ''duck and cover'' training as a response to a surprise attack.

4. *Report of the Secretary-General of the United Nations,* pp. 11–15 and 25–35.

2. The Cuban missile crisis led to the Fallout Shelter Program. Because ICBM technology shortened the presumed response to an hour or less and the development of the hydrogen bomb had greatly increased the destructiveness of bombs, sheltering was emphasized over evacuation. This program faced major resistance from a variety of sources (political, military,and public opinion). Despite the limits of such a program, particularly the lack of aid to targeted areas, this program continued for almost two decades.

3. Beginning in 1979, the very controversial Crisis Relocation Program was initiated. This program, still under development, presumes that a lead time of several days during a buildup of international tension will allow evacuation of major urban areas before any nuclear attack.

The idea of civil defense received early support from Brodie and more substantially from Kahn. Each discussed civil defense as a necessary component of maintaining a credible threat, particularly in defense of Europe. The Gaither committee had advocated a fallout shelter program in 1957, but the Eisenhower administration rejected the idea. Eisenhower was not sure that shelters would offer feasible protection, and he knew a national program financed by the federal government would be very expensive. Instead, with his National Shelter Program, Eisenhower made protection an individual responsibility. A government-funded shelter program was adopted in 1961 by the Kennedy administration and is analyzed by Kaufmann. A general historical overview is provided by Winkler. Recent Congressional hearings and reports include those by the Joint Committee on Defense Production and the Senate Committee on Banking, Housing and Urban Affairs. Scheer critiques the current civil defense organization, Federal Emergency Management Administration (FEMA) and the Crisis Relocation Program. Other recent critiques of current plans are found in Kerr and Leaning.

Antiballistic Missiles

A second countermeasure, and the one receiving the greatest technical emphasis, was the interest in an antiballistic missile (ABM) system. By 1966, the Soviet Union had developed a modest antiballistic system around Moscow. The following year, the United States announced it would construct a "thin" ABM system, purportedly in response to expected development of ICBM capacity by China. The first Strategic Arms Limitation Talks (SALT I) and subsequent political agreements have further weakened the development of such a program. When SALT I was signed in 1972, it included both the ABM Treaty and an interim agreement on ICBM and SLBM launchers. Initially, the ABM Treaty placed a limit of 200 ABM launchers for each side; this number was halved in 1974. Limits on ABM systems were designed to prevent either side from developing a first-strike capability, and more specifically to prevent effective damage limitation (especially from a retaliation), which would be necessary if a first strike were to be successful. Neither side would be likely to start a nuclear war if it knew it could not defend itself against retaliation.

Proponents and critics of ABM systems engaged in intense debate. Generally, the debate focused on three issues: (1) the nature of the arms race (whether it was an action/reaction phenomenon where a defensive race would begin in which countermeasures and counter-countermeasures would emerge); (2) the technical feasibility of ABM systems (whether proper missile accuracy and computer equipment could be developed); and (3) the strategic viability of ABM systems (whether they were a further deterrent or destabilizing influence). Debate over antiballistic systems dates back to the late 1940s and early 1950s, as described in Lapp and Freedman. Chayes and Weisner's study as well as that by Halperin (1971) provide important background information on the varying military and political perspectives toward defensive systems. Also relevant here are hearings of the Senate Foreign Relations Committee (1969).

The issue of an ABM system has also been widely debated among professional strategists and academicians. Three of the more balanced works on the subject are those by Coffey, Holst and Schneider, and the reader edited by Rabinowitz and Adams. Two articles, by Brennan and Dyson, stand out in the clarity of their advocacy of the ABM; lucid and convincing arguments against such a system are found in the work of Garwin and Bethe and of Stone.

Multiple Independently Targetable Reentry Vehicles

While ABM systems were generally viewed as attempts to reduce vulnerability to ICBMs, the development and deployment of multiple independently targetable reentry vehicles (MIRV) were seen as a response to ABM. Use of MIRVs significantly enhances the efficiency and mobility of the strategic weapons arsenal and can overwhelm a ground-based ABM system: three to ten warheads are placed on each land or sea-launched missile, and each warhead can be delivered to a separate target. The United States attained MIRV capacity on its Minuteman III missiles in 1970, with the first Poseidon submarine deployed the following year. The Soviet Union achieved MIRV capacity in 1975. The political debate over MIRV is reflected in the Senate Foreign Relations Committee report and the hearings of the House Committee on Foreign Affairs. Viewing MIRV as a case study in decision making are Greenwood and Tammen. The effects of MIRV on existing ICBM forces is discussed by York (1975) and by Davis and Schilling.

Counterforce Weapons

Since the early 1970s, nuclear weapons development has become increasingly qualitative. Refinements in the size (that is, yield) and accuracy of weapons systems have altered the potential use of the weapons. On the one hand, a direct relation holds between yield and accuracy: the greater the accuracy, the lower the yield necessary for the weapon to accomplish its mission. On the other hand, as yield decreases, so does the amount of damage beyond the point of detonation ("ground zero"). When intended for use against military targets (for example, missile silos), the yield of nuclear weapons has been decreased only as the accuracy has been increased.

The deployment of nuclear weapons that combine lower yield with greater accuracy is often associated with strategies for limited or protracted nuclear war. Such weapons are often termed counterforce weapons and are criticized by opponents as first-strike weapons—weapons that might be used first to destroy an opponent's nuclear weapons before the opponent could launch them. Concern over efforts at modernization of existing weapons and the development of others during the last few years has focused on this counterforce dimension. Aldridge's two recent works criticize the Trident, MX, and Cruise missile programs as increasing the possibilities of nuclear war by developing a first-strike capability. General background information is provided by Stockholm International Peace Research Institute (SIPRI, 1974), Tsipis, and the hearings of the House Committee on International Relations (1976). Specific discussions of the Trident are provided by Aldridge and the 1975 Government Accounting Office study. The importance of the cruise missile is discussed by Kennedy (1978) and Versbow.

Space Weapons

The development of new nuclear weapon systems may result from new technologies, perceived changes in an opponent's arsenal necessitating countermeasures, or changes in national security policy. Each of these factors may be seen in the March 1983 call by President Reagan for research, testing, and potential development of new defensive systems capable of intercepting and destroying Soviet ballistic missiles (the major component of the Soviet Triad). Such a system, the president argued, would replace deterrence as a strategy. The Reagan administration has proposed an initial five-year research and testing program that combines presumed technological possibilities with a resulting change in policy—a policy that would render nuclear weapons "impotent and obsolete." The emphasis upon defense is presented both as a countermeasure to presumed Soviet advances in the area and as a bold initiative that would significantly alter the arms race.

Two types of new defensive weapons are under review. On the one hand, space weapons (nuclear and nonnuclear) could be used to aid or replace an ABM system. The various laser, particle beam, X-ray, and related technologies currently under review are termed ballistic missile defense (BMD) systems, as they would be designed to destroy ICBMs and, perhaps, submarine ballistic missiles before they reach their targets. Such systems include space-based battleship stations with on-board lasers or reflecting systems relayed from earth.

Because a BMD system requires early detection of missile launching and even traditional responses to nuclear attack need as much advance warning of detonation as possible, satellites play an indispensable role. Hence, both United States and the Soviet Union are seeking countermeasures to intelligence-gathering and war-fighting satellites. This second area aims for the production of antisatellite, or ASAT, systems. Currently the Soviet Union uses SS-9 missiles to launch killer satellites that intercept targets in one to three orbits. The United States currently launches antisatellite weapons from F-15 jets eighteen miles in the atmosphere. Under the Reagan proposal, 112 new ASATs would be developed and carried on 56 modified F-15s.

As we discuss in Chapters 3 and 7, the current emphasis on space weapons must be viewed within the larger context of changing nuclear policies, national security strategy, and the theory of deterrence. While the debates surrounding consideration of space weapons are delineated in subsequent chapters, three recent works may be suggested for the reader interested in focused discussions of space weapons. The strongest support for development of a space defense system is provided by Graham. Arguing for withdrawal from the 1962 ABM treaty, this work emphasizes the importance of space defense systems against Soviet strategic arsenals. Ritchie offers a critique of laser weapon systems, but also includes descriptions of a wide variety of weapon systems. Jasani presents material on both BMD and ASAT systems and relates space technology to the arms race.

Nuclear Weapons of Other Nations

The development of nuclear weapons by the United States and the Soviet Union is often termed the *vertical* arms race; the term *horizontal* arms race refers to the acquisition of nuclear weapons by other countries. Although the United States and the Soviet Union currently possess approximately 95 percent of the global stockpiles of nuclear weapons, the arsenals of the People's Republic of China, the United Kingdom, and France are each capable of destroying almost any adversary. India, with its ''peaceful nuclear explosive experiment'' of 1974, is also a member of the ''nuclear club,'' technically speaking. The United Nations estimates that another twenty or more countries (including Israel, South Africa, Pakistan, and Iraq) have the capability to acquire a nuclear force.

This proliferation of nations with nuclear weapons is an area of growing concern for arms control advocates and opponents alike. Some of the major questions include: (1) Can or should nuclear power be expanded given the existing technology to transform spent fuel to weapon-grade levels? (2) Can or should the horizontal proliferation of nuclear weapon capacity be controlled? (3) Is nuclear terrorism a realistic possibility? (4) Does the spread of nuclear capacity to other countries increase the possibility of nuclear war? (5) Does the vertical proliferation of nuclear weapons by the United States and the Soviet Union increase the possibility of war between them? (6) Of what significance is the continued abrogation by both the United States and the Soviet Union of the Nonproliferation Treaty understanding that other nations would remain nonnuclear only if the superpowers worked seriously toward arms control and lessened their dependence on nuclear weapons? Dunn provides a popular discussion on the consequences of proliferation and the various mechanisms to deter further horizontal proliferation. *Nuclear Proliferation Factbook,* produced by the Senate Committee on Governmental Affairs (1977), is a practical introduction to the issues.

Comparative Nuclear Arsenals

Negotiations over arms control have generated a substantial literature comparing Soviet and American nuclear arsenals. Such comparisons generally focus on strategic weapons and are often used in discussions of parity, weapon vulnerability, modernization, and so on. A small number of works provide general trend data and descriptions of weapons systems. Polmar and the United Nations report provide a wide range of information. The recent work of Cochran, Arkin, and Hoenig discusses historical trends, changes, and analyses of current and future American nuclear weapons and policies. Useful annual comparisons of American and Soviet nuclear capacities are provided by the International Institute for Strategic Studies, the Stockholm International Peace Research Institute (1979) and Ruth Sivard.

Bibliography

The following bibliography includes references cited in the text and related suggested readings; references are organized according to the major divisions of the chapter. The full bibliographic reference for all sources cited in the text of this chapter may be found in either the annotated or unannotated bibliography corresponding to the division of the text in which the source is cited. Sources that are annotated are the most essential or significant.

Early Developments

Alperowitz, Gar. *Atomic Diplomacy: Hiroshima and Potsdam.* New York: Simon & Schuster, 1965.

An examination of the effects of the atomic bomb dropped on Hiroshima on American-Soviet relations. The author argues that the bomb dropped on Hiroshima was not necessarily for military purposes alone, but rather was intended equally as a political message to the Soviet Union that it would have to moderate expansionism because the United States possessed and would use atomic weapons.

Committee for the Compilation of Materials on Damage Caused by the Atomic Bomb. *Hiroshima and Nagasaki: The Physical, Medical and Social Effects of the Atomic Bombings.* Translated by Eisei Ishikawa and David L. Swain. New York: Basic Books, 1981.

The most authoritative compilation of data on the short and long-term consequences of the bombings of Hiroshima and Nagasaki. Includes technical, psychological, environmental, and medical studies.

Glasstone, Samuel and Philip J. Dolan. *The Effects of Nuclear Weapons.* Washington, D.C.: U.S. Department of Defense and U.S. Department of Energy, 1977.

The most authoritative explanation of nuclear explosions. While somewhat technical, its treatment of environmental, biological, and medical effects is essential for the researcher wanting to understand ecological effects.

Hewlett, Richard G. and Oscar E. Anderson. *The New World, 1939-1946.* Vol. I, *History of the U.S. Atomic Energy Commission.* University Park: Pennsylvania State Univ. Pr., 1962.

Part of the official history of the AEC, this book describes the Manhattan Project and the debate over whether control of atomic energy should be in military or civilian hands.

_____ and B. Duncan. *Atomic Shield, 1947-1952.* Vol. 2, *History of the U.S. Atomic Energy Commission.* University Park: Pennsylvania State Univ. Pr., 1970.

Part of the official history of the Atomic Energy Commission when it was responsible for developing both fission and fusion weapons and establishing atomic energy programs as power sources.

Smyth, Henry D. *Atomic Energy for Military Purposes: The Official Report on the Development of the Atomic Bomb.* Princeton, N.J.: Princeton Univ. Pr., 1945.

Part of the official history of the development of the atom bomb between the years 1939 and 1945. Describes the administrative view of the need and consequences of the atom bomb.

U.S. Strategic Bombing Survey. *The Effects of the Atom Bombs on Hiroshima and Nagasaki.* San Francisco: Gannon, 1973.

Part of the multivolume study on bombing, this represents the U.S. governmental view of the "experimental" use of atom bombs in Japan.

Armine, Michael. *The Great Decision: The Secret History of the Atomic Bomb.* New York: Putnam, 1959.

Baker, Paul R. *The Atomic Bomb: The Great Decision.* New York: Holt, Rinehart & Winston, 1968.

Barnet, Richard. *The Alliance: America, Europe, Japan.* Washington, D.C.: Institute for Policy Studies, 1984.

Beard, Edmund. *Developing the ICBM: A Study in Bureaucratic Politics.* New York: Columbia Univ. Pr., 1976.

Blackett, Patrick Maynard Stuart. *The Military and Political Consequences of Atomic Energy.* London: Turnstile, 1948.

Brown, Anthony and Charles McDonald, eds. *The Secret History of the Atomic Bomb.* New York: Dial, 1977.

Clark, Ronald W. *The Birth of the Bomb.* London: Phoenix House, 1961.

Compton, Arthur H. *Atomic Quest.* New York: Oxford Univ. Pr., 1956.

Davis, Nuel P. *Lawrence and Oppenheimer.* New York: Simon & Schuster, 1968.

Dyson, Freeman. *Disturbing the Universe: A Life in Science,* New York: Harper, 1979.

Feis, Herbert. *Japan Subdued: The Atomic Bomb and the End of the War in the Pacific.* Princeton, N.J.: Princeton Univ. Pr., 1961.

Fogelman, Edwin, ed. *Hiroshima: The Decision to Drop the Bomb.* New York: Scribner, 1974.

Giovannetti, Len and Fred Freed. *The Decision to Drop the Bomb.* New York: Coward, 1965.

Groves, Leslie. *Now It Can Be Told.* New York: Harper, 1962.

Hersey, John. *Hiroshima.* New York: Knopf, 1946.

Hiroshima/Nagasaki: August, 1945. New York: Museum of Modern Art, n.d. Film.

Kaufmann, William, ed. *Military Policy and National Security.* Princeton, N.J.: Princeton Univ. Pr., 1956.

Kennan, George F. *Russia, the Atom, and the West.* New York: Harper, 1958.

Kennedy, Robert F. *Thirteen Days: A Memoir of the Cuban Missile Crisis.* New York: Norton, 1969.

Lamont, Lansing. *Day of Trinity.* New York: Knopf, 1946.

Laurence, William L. *Dawn over Zero.* New York: Knopf, 1946.

_____. *Men and Atoms: The Uses and the Future of Atomic Energy.* New York: Simon & Schuster, 1959.

Lens, Sidney. *Day before Doomsday: An Anatomy of the Nuclear Arms Race.* Garden City, N.J.: Doubleday, 1977.

Lifton, Robert J. *Death in Life: Survivors of Hiroshima.* New York: Random, 1968.

Lilienthal, David E. *The Journals of David E. Lilienthal.* New York: Harper, 1964.

Mandelbaum, Michael. *The Nuclear Question: The United States and Nuclear Weapons, 1946-1976.* Cambridge: Cambridge Univ. Pr., 1979.

Moss, Norman. *Men Who Play God: The Story of the H-Bomb and How the World Came to Live with It.* New York: Harper, 1968.

Oughterson, Ashley W. and Warren Shields, eds. *Medial Effects of the Atomic Bomb in Japan.* New York: McGraw-Hill, 1956.

SANE Education Fund. *Shadows of the Nuclear Age.* Philadelphia: SANE Education Fund, 1981. Radio tapes.

Schilling, Warner R. "The H-Bomb Decision: How to Decide without Actually Choosing." *Political Science Quarterly* 76 (March 1961):24–46.

Sherwin, Martin J. *A World Destroyed: The Atom and the Grand Alliance.* New York: Knopf, 1975.

Stimson, Henry Lewis and McGeorge Bundy. *On Active Service in Peace and War.* London: Hutchinson, 1948.

Teller, Edward with Allen Brown. *The Legacy of Hiroshima.* Garden City, N.J.: Doubleday, 1962.

Time-Life Films. *Building of the Bomb.* New York: Time-Life Films, 1975. Film.

York, Herbert F. *Road to Oblivion: A Participant's View of the Arms Race.* New York: Simon & Schuster, 1970.

————. *The Advisors: Oppenheimer, Teller, and the Superbomb.* San Francisco: Freeman, 1976.

Development of Strategic and Nonstrategic Weapons

Etzold, Thomas H. and J. L. Gaddis. *Containment: Documents on American Policy and Strategy 1945–1950.* New York: Columbia Univ. Pr., 1978.

> Excellent collection of internal reports, emergence of Defense Department, NATO, and military strategy toward Soviet Union. Of particular interest is the treatment of strategic considerations toward the atomic bomb immediately after World War II.

Freedman, Lawrence. *The Evolution of Nuclear Strategy.* New York: St. Martin's, 1981.

> Presents a detailed analysis of American nuclear strategy from Hiroshima to the present. Argues that initial bombing in Japan was seen as consistent with a policy of strategic bombing in which the focus is on demoralization of civilian population. Argues that debates over strategy are cyclical rather than linear: we continue to build new programs and new policies, none of which successfully resolve basic issues.

Halperin, Morton. *Contemporary Military Strategy.* Boston: Little, 1967.

> A short history and analysis of American military strategy from the end of World War II to the mid-1960s. Strength of book is in the description of the ways in which military planners defined international political relations and their effects on strategy.

Brenner, Michael. "Tactical Nuclear Strategy and European Defense: A Critical Reappraisal." *International Affairs* 51, no. 1 (January 1975): 23–42.

Dyer, Philip W. "Tactical Nuclear Weapons and Deterrence in Europe." *Political Science Quarterly* 92 no. 2 (Summer 1977): 245–58.

Gray, Colin S. "Deterrence and Defense in Europe: Revising NATO's Theatre Nuclear Posture." *Strategic Review* 3, no. 3 (1975): 41-51.

Security Resources Panel of the Scientific Advisory Committee of the Office of Defense Mobilization. *Deterrence and Survival in the Nuclear Age* (The Gaither Report). Washington, D.C.: Govt. Print. Off., November 1957.

Technological Capabilities Panel of the Scientific Advisory Committee of the Office of Defense Mobilization. *Meeting the Threat of Surprise Attack* (The Killian Report). Washington, D.C.: Govt. Print. Off., February 14, 1955.

U.S. Congress. Congressional Budget Office. *Planning U.S. General Purpose Forces: The Theater Nuclear Forces.* 95th Cong., 1st session. Washington, D.C.: Govt. Print. Off., April 1977.

U.S. Department of Defense. *FY19— Department of Defense Program for Research, Development, and Acquisitions.* Washington, D.C.: Govt. Print. Off.

Van Cleave, William R. and S. T. Cohen. *Tactical Nuclear Weapons: An Examination of the Issues.* New York: Crane, Russak, 1978.

Nuclear Weapons in the Modern Era

Aldridge, Robert C. *First Strike! The Pentagon's Strategy for Nuclear War.* Boston: South End, 1983.

This is a highly critical examination of new weapons systems (Trident, Cruise, MX) that increase the chance of nuclear war. While detailed, the book provides a one-time insider's view of military dependence on nuclear weapons.

Chayes, Abram and Jerome Weisner, eds. *ABM: An Evaluation of the Decision to Deploy an Anti-Ballistic Missile System.* New York: Harper, 1969.

Edited work critical of the decision to deploy the ABM system; argues that it was militarily unsound, politically unwise, and technically infeasible.

Graham, Daniel. *High Frontier: A New National Strategy.* Washington, D.C.: Heritage Foundation, 1983.

Retired general argues for American superiority; describes a space network that would purportedly be capable of destroying Soviet strategic weapons. While highly expensive, such a system is, he argues, a primary means for achieving dominance in the arms race.

Greenwood, Ted. *Making the MIRV: A Study of Defense Decision Making.* Cambridge, Mass.: Ballinger, 1975.

Analysis of the technical, political, and bureaucratic factors affecting the development of the MIRV system. The author compares the roles of the intelligence community (about whether the Soviet Union was developing a similar system) with the relationship between military strategy and policy. Provides a discussion of the resistance to the program and extrapolates a series of propositions relevant to weapon development.

Jasani, Bhupendra. *Outer Space: A New Dimension of the Arms Race.* Stockholm: Oelgeschlager, Gunn, & Hain, 1982.

A series of fifteen papers focusing on space technology and arms control. Issues covered include monitoring, verification, United States vs. Soviet strategies toward space, and attempts at controlling or banning weapons in space. While technical in orientation, it remains a useful introduction to the types of space technology of relevance to nuclear weapons.

Kahn, Herman. *On Thermonuclear War.* Princeton, N.J.: Princeton Univ. Pr., 1960.

Classic elaboration of theory of deterrence and nuclear war. He discusses the differences between conventional and nuclear war, and between first and second strike capabilities.

Polmar, Norman. *Strategic Weapons: An Introduction.* New York: Crane, Russak, 1975.

General reference work on nuclear weapons from 1945 to the mid-1970s. Includes trend data on deployment of weapons.

Report of the Secretary-General of the United Nations. *Nuclear Weapons.* Brookline, Mass.: Autumn Pr., 1982.

Official report to the General Assembly of the United Nations. Neither the United States nor the Soviet Union participated in preparation of report. The focus is on nuclear weapons, doctrines regarding their use, security implications of continued arms race.

Stockholm International Peace Research Institute (SIPRI). *World Armament and Disarmament: SIPRI Yearbook, 1978.* New York: Crane, Russak, 1979.

Annual analysis of weapons development (nuclear and nonnuclear) with supporting technical data. Also discussed are efforts for arms control and disarmament. A shorter version of book is available under the title *The Arms Race and Arms Control.*

Aldridge, Robert C. *Counterforce Syndrome: A Guide to U.S. Nuclear Weapons and Strategic Doctrine.* Washington, D.C.: Institute for Policy Studies, 1981.

Brennan, Donald G. "The Case for Missile Defense." *Foreign Affairs* 47 (April 1969): 433–48.

Brodie, Bernard. *Strategy in the Missile Age.* Princeton, N.J.: Princeton Univ. Pr., 1959.

Cochran, Thomas B., William M. Arkin, and Milton M. Hoenig. *Nuclear Weapons Databook.* Vol. I, *U.S. Nuclear Forces and Capabilities.* Cambridge, Mass.: Ballinger, 1984.

Coffey, Joseph I. "The Anti-Ballistic Missile Debate." *Foreign Affairs* 45 (April 1967): 403–13.

Constant, James N. *Fundamentals of Strategic Weapons.* 2 vols. Hague: Martinus Nijhoff, 1983.

Davis, Lynn Etheridge and Warner R. Schilling. "All You Ever Wanted to Know about MIRV and ICBM Calculations but Were Not Cleared to Ask." *Journal of Conflict Resolution* 17, no. 2 (June 1973): 207–42.

Dunn, Lewis. *Controlling the Bomb: Nuclear Proliferation in the 1980s.* New Haven, Conn.: Yale Univ. Pr., 1982.

Dyson, Freeman. "A Case for Missile Defense." *Bulletin of Atomic Scientists* 25 (April 1969): 431–33.

Garwin, Richard L. and Hans A. Bethe. "Anti-Ballistic Missile Systems." *Scientific American* 218, no.3 (March 1968): 21–31.

Gray, Colin S. *The MX, ICBM, and National Security.* New York: Praeger, 1981.

Halperin, Morton. *The Decision to Deploy ABM: Bureaucratic and Domestic Politics in the Johnson Administration.* Washington, D.C.: Brookings, 1971.

Holst, Johan Jorgen and William Schneider. *Why ABM? Policy Issues in the Missile Defense Controversy.* Elmsford, N.Y.: Pergamon, 1969.

International Institute for Strategic Studies. *Strategic Survey, 1978.* Boulder, Colo.: Westview, 1978.

Kaufmann, William. *The McNamara Strategy.* New York: Harper, 1964.

Kennedy, Robert. "The Cruise Missile and the Strategic Balance." *Parameters* (March 1978): 62–72.

Kerr, Thomas J. *Civil Defense in the United States: Bandaid for a Holocaust?* Boulder, Colo.: Westview, 1983.

Lapp, Ralph. *The Weapons Culture.* New York: Norton, 1968.

Leaning, Jennifer. *Civil Defense in the Nuclear Age: What Purpose Does It Serve and What Survival Does It Promise?* Cambridge, Mass.: Physicians for Social Responsibility, 1982.

Rabinowitch, Eugene and Ruth Adams, eds. *Debate: The Antiballistic Missile.* Chicago: Bulletin of Atomic Scientists, 1967.

Ritchie, David. *Spacewar.* New York: Atheneum, 1983.

Scheer, Robert. *With Enough Shovels.* New York: Knopf, 1982.

Sivard, Ruth. *World Military and Social Expenditures, 1974– .* Washington, D.C.: World Priorities, 1975.

Stockholm International Peace Research Institute. *Offensive Missiles: Stockholm Paper 5.* Stockholm: SIPRI, 1974.

_____. *Arms Uncontrolled.* Prepared by Frank Barnaby and Ronald Huisken. Cambridge: Harvard Univ. Pr., 1975.

Stone, Jeremy. *The Case against Missile Defense.* London: Institute for Strategic Studies, 1968.

Tammen, Ronald L. *MIRV and the Arms Race: An Interpretation of Defense Strategy,* New York: Praeger, 1973.

Tsipis, Kosta. "The Accuracy of Strategic Missiles." *Scientific American* 238, no. 1 (July 1975): 14–23.

U.S. Congress. House. Committee on Foreign Affairs. Hearings. *Diplomatic and Strategic Impact of Multiple Warhead Missiles.* Washington, D.C.: Govt. Print. Off., 1969.

_____. Committee on International Relations. *Nuclear Proliferation Factbook.* 95th Cong. 1st sess. Washington, D.C.: Govt. Print. Off., September 1977.

_____. Subcommittee on International Security. Hearings. *First Use of Nuclear Weapons: Preserving Responsible Control.* 94th Cong. 2nd sess. Washington, D.C.: Govt. Print. Off., March 1976.

_____. Joint Committee on Defense Production. Hearings. *Civil Preparedness and Limited Nuclear War.* 94th Cong., 2nd sess. Washington, D.C.: Govt. Print. Off., April 1976.

_____. Senate. Committee on Foreign Relations. Hearings. *Strategic and Foreign Policy Implications of ABM System.* Washington, D.C.: Govt. Print. Off., 1969.

_____. Subcommittee on Arms Control, International Law and Organization. *ABM, MIRV, SALT, and the Nuclear Arms Race.* 91st Cong., 2nd sess. Washington, D.C.: Govt. Print. Off., 1970.

_____. Committee on Banking, Housing, and Urban Affairs. Hearings. *Civil Defense.* 95th Cong., 2nd sess., Washington, D.C.: Govt. Print. Off., 1979.

U.S. General Accounting Office. *Trident and SSBN-X Systems.* Staff Report. Washington, D.C.: Govt. Print. Off., February 1975.

Versbow, Alexander. "The Cruise Missile: The End of Arms Control?" *Foreign Affairs* 55, no. 1 (October 1976): 133-46.

Winkler, Allan M. "A 40-Year History of Civil Defense." *Bulletin of Atomic Scientists* 40 (June/July 1984): 16–23.

York, Herbert F., ed. *Arms Control.* San Francisco: Freeman, 1973.

_____. "The Origins of MIRV." In *Dynamics of the Arms Race,* edited by David Carlton and Carol Schaerf. New York: Wiley, 1975.

CHAPTER 3

Nuclear Policies

The rise of nuclear weapons has provoked new policy considerations: Should nuclear weapons be treated as primarily a bigger, more destructive weapon or as a weapon so different and significant that traditional military strategy is called into question? What is the relationship between conventional and nuclear weapons in war fighting? Is it possible or plausible to attempt passive or active defense measures to nuclear weapons? What are the ramifications of devising countervalue and counterforce targeting options? More generally, is the traditional dependence on military strength viable as a means of resolving interstate conflict in an era of nuclear parity between the superpowers?

These and related issues have become paramount concerns over the past four decades. They have been and will continue to be debated in public and private forums by politicians, strategists, and military planners as well as by those traditionally viewed as outside the national security decision-making process (for example, private citizens, clergy, and scientists). Current debates over issues as varied as modernization of strategic forces (such as the MX and Trident), relations with NATO (such as deployment of Euromissiles, the firebreak between nuclear and conventional weapons, ''no first use''), deterrence (such as assured destruction vs. flexible response), and the recent emphasis on defensive weapon systems (such as BMDs, ASATs) are imbedded within the larger nexus of political and military policy.

Yet what constitutes nuclear policy? More specifically, how does the average citizen begin to understand policies of relevance to nuclear weapons and war? Is there a clear articulation of such policies and, if so, is it publicly accessible? In general, the answers to these questions are difficult to ascertain for a number of reasons.

First, there are a wide variety of relevant policy dimensions, not all of which are consistent. Ball, for example, distinguishes between declaratory policy (public pronouncements of official rationales); force development policy (concerning size and capability of existing and proposed weapon systems); policy to guide arms control negotiations, operational policy (for example, concerning patrol practices of nuclear submarines); and force employment or action policy (such as plans concerning the use of nuclear forces). Inconsistencies between declaratory and force employment policy, for example, may lead to debates having little relevance to actual war plans.

Second, while information on declaratory, force deployment, and arms control policy may be publicly accessible, operational and force employment policy is normally secret. Recently declassified documents concerning force employment policy in the period following World War II suggest, for example, a wide disparity between the public pronouncements and actual targeting plans.

Third, the public articulation of policy is directed toward different audiences. With audiences as distinct as Congress, the American public, political allies, and, importantly, adversaries, policy articulation may intentionally be sporadically vague, specific, or even disingenuous.

Fourth, nuclear policies represent fragmented decision making. This fragmentation reflects both the multiple sources of policy development and the competing perspectives of military planners, diplomats, and civilian strategists.

Fifth, since the late 1950s policy formulation (and public reaction to it) has been greatly influenced by specialists in strategic affairs and deterrence theory. This has often led to terminology, assumptions, and arguments quite alien to public understanding. References to "second-strike capability" or "escalation control" are often viewed by the public as too technical and beyond common understanding.

Despite these barriers, we argue that it is possible and important for the interested person to more fully understand nuclear policy. At any given historical or current period, there is a general public awareness of what constitutes nuclear policy. While this general awareness may be imprecise, imperfect and, in certain cases, wrong, it remains true that a general sense of policy exists. Moreover, our understanding and interpretation of *past* policy are altered as various documents become published or declassified. Finally, a fairly substantial number of writings have recently been published oriented toward nonprofessional audiences that trace the evolution of and debates over policy.

Reference and research material for those seeking a deeper analysis of policy includes (1) official documents and speeches by presidents, secretaries of state and defense, directors of the Central Intelligence Agency, and other agency heads; (2) Congressional hearings and legislation on nuclear forces, strategies, and arms control agreements; (3) personal memoirs and biographies of important decision makers and strategists (for example, Truman, Kissinger, Kennan, McNamara); and (4) academic research, particularly analyses of foreign policy and international relations. Materials most likely to remain classified for substantial periods of time include military-initiated or sponsored research, secret hearings before Congress, and internal documents, including those of the National Security Council (see the Appendix for descriptions of many of these sources).

Despite its diversity, the history of nuclear policy suggests three periods, generally recognized but variously defined, during which the dominant policies of the United States were shaped by the actual or perceived levels of Soviet technology and policy. This chapter reviews (1) the early Cold War period and the policy of containment (the period from 1945, when the United States had exclusive possession of nuclear weapons, to the end of its clear monopoly); (2) the policies of massive retaliation and mutually assured destruction (MAD), which began in the early 1950s and continued through at least the 1960s, during which the United States clearly maintained nuclear superiority in weapons; and (3) the period of the 1970s and 1980s for which consensus is lacking over the relative position of the nuclear arsenals of the Soviet Union and the United States and the status of nuclear policy. In general, regardless of the various policy names used and the distinct emphasis given by successive presidents, two characteristics have remained constant in U.S. nuclear policy: rejection of what has come to be termed a no first use posture and advocacy for a position that, technically, is termed flexible response—whether and at what levels use of conventional and nuclear weapons will occur is left open so as to increase options and, it is hoped, deterrence.

We will trace in detail and in separate sections these three periods as they occur in American policy and, in a final section, suggest readings for those interested in the nuclear policies of other nations. Readers wishing general treatments that focus on the evolution of policy should consult Freedman, Mandlebaum, the anthology edited by Head and Rokke, and the relevant volumes of the series produced under the direction of Ferrell, *American Secretaries of State and Their Diplomacy*. At even broader levels, good treatments of foreign policy (especially on the United States and the Soviet Union) include Ulam and Spanier.

The Early Cold War Period and Containment

Interest in the early development of nuclear weapons policy has increased recently, particularly as documents of the period have been declassified and made available for analysis. These documents, from the U.S. State Department series *Foreign Relations of the United States* as well from the collection presented by Etzold and Gaddis, when combined with the works of Alperowitz, Freedman, and Yergin, suggest that early nuclear policy (1) emerged as a continuation of policies developed during World War II; (2) served as a response to the postwar perception of the Soviet Union; and (3) developed as a result of the failure of the Baruch plan and the shift to such regional alliances as NATO.

Continuity

The first policy on nuclear weapons was a policy of use. During World War II, debate centered on the ways to use the atom bomb (see Chapter 2). The atomic bombing of Hiroshima and Nagasaki continued the policy of strategic bombing of civilian rather than military targets (intended to weaken the psychological resolve of an enemy). Debate continues today on whether the decision to bomb Japan was only to force the surrender of Japan or was primarily a message to the Soviet Union. The literature on this topic includes Sherwin, Yergin, Bernstein, and Alperowitz. Internal security studies and planning documents of the late World War II period and early postwar years suggesting continuity are provided by Masters and Way, Etzold and Gaddis, and Quester. (See especially the Compton, Jeffries, Spaatz, and Harmon reports.) These works discuss choice of targets, using the bomb as part of a surprise attack or as a means of last resort, and interservice rivalry over the bomb.

Perception of the Soviet Union

The policy guiding use of nuclear weapons, emergent by 1948, came to be associated with the policy of containment as outlined by George Kennan. In a series of communications submitted while he served first as a deputy to the American ambassador to the Soviet Union and subsequently as director of the Policy Planning Staff at the Department of State, Kennan argued that the United States should adopt policies containing the Soviet Union's expansionist actions in Eastern and Western Europe. The notion of containment, which Kennan intended to refer to political means, came to be incorporated as a military policy by the Truman administration. In his *Memoirs*, Kennan stresses this difference between his intent and administrative policy.

The Etzold and Gaddis work is indispensable for understanding Kennan's influence on policy. This collection includes various Kennan writings and National Security Council documents, notably the recently declassified NSC-30, which in 1948 calls for a place for atomic weapons in U.S. military planning, and NSC-68, which in 1950, a year after the Soviet Union's first atom bomb test, led President Truman to order production of the hydrogen bomb and make other conventional and nuclear additions to the United States forces. These and other documents make clear that containment of the Soviet Union could involve use of atomic weapons and that such use could be activated prior to Soviet use of their atomic weapons.

The Baruch Plan and NATO

In 1946, the United States proposed the Baruch plan whereby all fissionable material would be placed under international control (the draft of the text of this plan can be found in *Documents on Disarmament*). Under this plan, no nation could possess nuclear weapons; verification would be provided through inspections and the applications of sanctions. The Soviet Union subsequently rejected the plan, arguing that the provision for onsite inspections violated national sovereignty. Its alternative proposal, which included no inspection but advocated the prohibition of nuclear weapons and destruction of existing stock, is described in Molotov. Also useful on this topic are the recommendations of the United Nations Atomic Energy Commission (U.S. Dept. of State, 1949), Baruch, and Leiberman.

Given the veto power held by the five permanent members of the United Nations Security Council, the United States abandoned hope for a collective agreement on nuclear weapons and moved toward the development of regional alliances. The North Atlantic Treaty Organization (NATO), formed in 1949, promised American conventional and nuclear forces in Europe as a deterrent to Soviet expansion.

The three elements of continuity with World War II strategic bombing, the failed attempts at international control, and the adaptation of a containment perspective toward the Soviet Union should be viewed within the context of American monopoly over nuclear weapons. As long as the United States alone had a nuclear capacity, the alternatives for policy and the particular strategies for military use reflected the extension of past bombing ideas with discussions of the potential advantages represented by the atomic bomb. However, as Brodie (1946) noted in an early work, the eventual loss of monopoly status would require important shifts. Not only would the ability to retaliate after a nuclear attack become central, but the very essence of nuclear weapons themselves would have to be expanded to include deterring the outbreak of a war.

From Massive Retaliation to Mutual Assured Destruction

The American monopoly over nuclear weapons strongly shaped the early policies concerning their use. The period of containment was characterized by a debate between those arguing for an offensive function and those viewing the atom bomb as a defensive tool, either as a weapon to be used as a last resort or as a means of protecting Western Europe from the Soviet Union. The policy shifts characterizing the mid-1950s and 1960s were not so much sudden reversals of direction as recognition of the Soviet Union's emergence as a nuclear power and mutual concern with vulnerability. Attacks on the credibility of existing policy, changing emphasis on offensive and defensive postures, tensions among military, civilian, and political orientations toward nuclear weapons, and the public and private stances of presidents and their advisors all reflect the shift in policy.

Official and scholarly literature suggests that American nuclear policy went through three phases between the enunciation of massive retaliation in 1954 and the call for strategic arms limitation talks in 1967: (1) the affirmation of an explicitly offensive military posture; (2) the rejection of an all-out nuclear attack as the only operational option for use; and (3) the focus on assured destruction in a major nuclear exchange. What began as a critique of containment in the early 1950s ended with a critique of massive retaliation in the 1960s.

Massive Retaliation

Despite then-secret National Security Council documents that indicate U.S. nuclear policy included first-strike options, the doctrine of containment was widely understood as an essentially defensive and responsive policy. If the United States did respond, for example to a Soviet invasion of Western Europe, it could initiate nuclear retaliation, even massive retaliation with nuclear weapons. However, the stress was on more measured responses. This policy came under increasing criticism with the Eisenhower administration. Critics argued that containment placed the United States in a reactive position, not able to use its weapons until Soviet actions precipitated an American response; moreover, the containment policy was seen as ignoring the relevance of tactical or theater nuclear weapons as a means of blocking a Soviet invasion of Western Europe. This latter criticism is more fully stated in the Schilling, Hammond, and Snyder work.

Massive retaliation, as a publicly stated policy, emerged in 1954 in a speech by John Foster Dulles. Popular interpretation was that the speech indicated a major shift in policy, although the actual policy (as outlined in National Security Council document 162/2 and the subsequently published version of Dulles's speech) indicated more of a change in emphasis than direction, that is, that nuclear weapons were among the available weapons to be used in response to Soviet aggression. The published version of Dulles's speech is reprinted in Head and Rokke; the NSC document can be found in *The Pentagon Papers*.

Understood popularly as a policy that called for immediate, massive nuclear retaliation to almost any Soviet action and that allowed the American arsenal to be used at its own discretion, massive retaliation led to debate over the size and composition of nuclear forces. It also led to increasing public concern, particularly in relation to the growing awareness of the greater lethality represented by the hydrogen bomb and to the radiation released in above-ground tests of newly developed weapons.

These public concerns are presented by both Divine and Wasserman et al.

During the late 1950s, criticism of the policy of massive retaliation grew stronger. Particularly important are the essays included in Kaufmann. This work argues that massive retaliation is seen as overly rigid with two potential negative alternatives—holocaust or humiliation, suicide or surrender. Adding to this debate is Kecskemeti's provocative text and the resulting legislation (as treated in King). Because of the prospect of mutual nuclear destruction, Kecskemeti argues against pursuit of unconditional surrender and for limited objectives in war. His book was misunderstood by some in Congress as a study on when and how the United States should surrender. As a result, an amendment was placed on a military appropriations bill that prohibits governmental funding for any studies on when the United States should surrender.[1] Other works, including Osgood's work on limited war and that of Kissinger (1957), received considerable attention in their efforts to suggest policy options that could pass between the horns of the dilemma of suicide or surrender.

Flexible Response

Momentum against the policy of massive retaliation grew in the late 1950s and early 1960s, partially in response to Soviet advances in ICBM technology and the launching of Sputnik and partially in relation to the alleged missile gap issue that surfaced during the presidential campaign of 1960. The work of Kennedy and Kaufmann's study of the early defense policy in the Kennedy administration (1964) provide important discussions of the move away from massive retaliation.

This shift was made operational in 1962 with the city avoidance strategy announced in Defense Secretary McNamara's "Defense Arrangements of the North Atlantic Community." The move away form targeting cities (countervalue targets) toward targeting an enemy's military forces (counterforce targeting) offered an alternative to the early policy of potential all-out attack. Attacks on cities are associated with total war—the complete destruction of a nation, including its people and economy. Strikes against military targets are associated with more limited objectives—a nation's people, to as large a degree as possible, and sometimes its economy are spared. Given the possibility of escalation to full-scale nuclear war, political leaders desired nuclear options that might not necessarily be interpreted as a prelude to an all-out attack and that, as a result, would facilitate similar responses if retaliation occurred. Useful discussions of this shift are found in McNamara (1968), Ball, and Enthoven and Smith.

---

1. Lawrence Freedman, *The Evolution of Nuclear Strategy* (New York: St. Martin's, 1981), p. 97.

Flexible response altered not only targeting strategy but also the defensive options available in response to any particular Soviet action. As commonly understood, flexible response came to mean consideration of various nuclear options short of all-out attack. It included implementation into NATO policy of a first-strike option (use of nuclear weapons to stop a conventional Soviet invasion of Western Europe).

Mutual Assured Destruction (MAD)

While the policy of flexible response has been the generally accepted policy of the United States since the early 1960s, important shifts in emphasis emerged in relation to assured destruction and the arms race as an action-reaction spiral. By 1964, assured destruction became a focal element of American nuclear strategy, in part to correct possible Soviet suspicion that the city avoidance policy and flexible response strategy were actually an implicit first-strike strategy. The emphasis on assured destruction stressed the mutual vulnerability of U.S. and Soviet populations and sought to preserve the "balance of terror" in which each side held hostage the other's population.

The problem was the action-reaction strategy of the arms race. Because of Soviet advances in weapon technology, both the United States and the Soviet Union were at a point of mutually assured destruction: no matter which side struck first, the other side could respond with a second strike at assured destruction levels. Both countries' forces had grown to increasingly strong and comparable levels. Further stockpiling of weapons came to be seen by many as counterproductive; instead, beginning in 1967, serious attention turned to the idea of limiting strategic arms and to debate over the utility of building an antiballistic missile system (ABM).

As responses to excessive stockpiling, efforts at limiting both strategic arms and antiballistic missiles were designed to enhance stability and reduce the danger of nuclear war through either the sheer momentum of continued arms buildup or the false sense that an attacker with an extensive ABM system could emerge relatively unscathed in a nuclear exchange. The interim agreement and ABM Treaty of SALT I were the results of this approach to enhancing stability (for more details on the strategic debate on these views, see Chapter 7).

Rejecting MAD

The Nixon, Ford, Carter, and Reagan administrations have moved away from a declaratory policy emphasis on assured destruction and toward a wide series of policies implicit in the strategy of flexible response: consideration of limited nuclear war, support for construction of weapons systems constituting the Triad strategic force, vacillating debate over ABM systems, and an emphasis upon "prevailing" in a protracted nuclear war.[2] The operational implications of these policy shifts are reflected in the development of the single integrated operational plan (SIOP). The third section of Chapter 4 discusses SIOP in detail. Public documentation of these policy shifts is found in the Schlesinger doctrine (pursued during the Nixon and Ford administrations) and Presidential Directive 59 (pursued during the Carter and Reagan administrations).

Concurrent with the emerging rejection of MAD as the primary policy orientation were the efforts at arms control, particularly the Strategic Arms Limitation Talks and their corresponding treaties. The agreements negotiated during the 1970s did, at least temporarily, reduce the interest in antiballistic defense systems and stabilize the arms race, but did not significantly reduce the size of existing arsenals or block development of weapon systems consistent with a shift in policy (for example, MIRV and cruise missiles).

Additionally, controversy continues over alleged violations of existing SALT agreements. Responses to presumed violations, depending on whether policy aims at parity or pushes for superiority, act as an impetus to ignore (that is facilitate more weapons) or reject the limits of existing policy agreements. It should be noted that while each side has publicly accused the other of violating SALT I (which is ratified) and SALT II (which is not ratified and has verbal rather than legal commitment), each alleged violation brought before the Standing Consultative Committee (a bilateral group emerging out of SALT I to resolve allegations of treaty violations) has been resolved or dropped.

A further and related development during the 1970s involved the American response to the Soviet build-up or modernization of its conventional and nuclear forces. Three missiles (the SS-17, SS-18, and SS-19) were produced, submarines were improved, (particularly the Typhoon and Delta class) and the Backfire bomber was deployed. Such developments intensified strategic debate in the United States and Europe and became a further rationale for the subsequent American arms buildup.

Out of the shift in policy emphasis, the attack on arms control, and fears concerning the Soviet modernization program came an emphasis on developing new U.S. weapon systems. Flexible response, particularly with its focus on counterforce rather than countervalue targeting, facilitates a move away from emphasis on all-out nuclear war and toward limited objectives that require more diverse and accurate weapons. As Secretary McNamara pointed out in the late 1960s, the strategic arsenals of both the United States and the Soviet Union were more than sufficient (then several hundred and currently about ten thousand strategic weapons on each side) for assured destruction, even in response to a first

2. For a detailed treatment of these shifts, see Robert Jervis, *The Illogic of American Nuclear Strategy* (Ithaca, N.Y.: Cornell Univ. Pr., 1984).

strike. Even in the early years, these strategic arsenals were of sufficient size to be used both for countervalue and counterforce options (for example, against Soviet conventional and nuclear forces as well as against the Soviet leadership and economy). Nevertheless, since the early emphasis was on countervalue options, it was possible for governments to be relatively content with the number of nuclear weapons.

Discontent came as other policy options received greater stress. Ironically, plans to fight limited and protracted wars call for more and improved weapons, though this tendency toward new weapons development has been coupled with a decrease in megatonnage for the strategic arsenal. More are required because they are being integrated into systems designed for several (rather than a single) strategic options. While political debate has focused on whether such a decrease in megatonnage signifies a loss of potential military strength, it would appear to reflect more realistically a lead in qualitative technology and to be consistent with a counterforce strategy (as smaller, more accurate weapons give a decided edge over weapons characterized by low accuracy but high megatonnage—as found in the Soviet strategic arsenal). Such a shift in weapon development and policy development does not, of course, reduce the danger that a large-scale nuclear war might arise from accident or escalation (see Chapter 4).

There is widespread debate today over whether current policy in the Reagan administration should be viewed as one of flexible response or some categorization more consistent with the ideas expressed by the Strategic Defense Initiative. Given the fact that the current administration has reemphasized arms buildup, expanded study on protracted and limited war, and, until quite recently, agreed to abide by past SALT agreements, it seems valid to characterize current policy as one of flexible response within a countervailing strategy: a doctrine that rejects the idea of assured destruction (deterrence) and argues that the United States will use whatever force necessary, at every level of conflict, to prevail in a conflict (increased war fighting capacity). This countervailing policy remains operational today, facilitating the development of new and updated weapons systems and improved command, control, communication, and intelligence (C3I) capacity. At the same time, increased attention has been given to a theoretical possibility—an alternative policy focusing on defense. While this suggests the potential of an important policy shift, it remains at this point only a potentiality, at least in terms of force employment policy.

Table 2 summarizes the major patterns discussed in Chapters 2 and 3. The table describes the varying policies and weapons dominant at three periods of time: (1) nuclear monopoly (1945–52); (2) nuclear superiority (1952–69); and (3) nuclear parity (1970 to present). No specific distinction is made between the last two periods of time, for assured destruction has remained the stated policy during both periods.

Table 2.
The Evolution of Nuclear Weapons and Policies

	Nuclear Monopoly (1945–1952)		Nuclear Superiority (1953–1969)		Nuclear Parity (1970–present)	
Nuclear policy	Containment	Massive retaliation	Flexible response	Assured destruction	Assured destruction	Flexible response; limited war
Weapons and types of new weapons	Atomic and strategic (bombers)	Strategic and tactical; ICBM; theater	Strategic; TRIAD and tactical	Strategic TRIAD and tactical ABM	Strategic TRIAD and tactical MIRV	TRIAD and tactical BMD and ASAT
Use of weapons: offensive or defensive; first or second strike; first use	Defensive; second strike	Offensive; first strike	Defensive; first use	Defensive; second strike	Defensive; second strike	Offensive; second strike; first use
Targeting	Countervalue (cities)	Countervalue (cities)	Counterforce (city avoidance)	Countervalue	Countervalue	Counterforce; countervalue
Perception of Soviet Union	Aggressive, strong, conventional forces (no bomb)	Strong conventional forces; weak nuclear forces	Strong conventional forces; weak nuclear forces	Superior conventional forces in Europe, strong nuclear forces	Superior conventional forces in Europe, strong nuclear forces	Superior conventional forces in Europe, strong nuclear forces
Policy concerns: credible policy; flexible policy; system vulnerable	Credible; not flexible; not vulnerable	Not credible; not flexible; vulnerable	Credible; flexible; vulnerable?	Not credible; not flexible; vulnerable	Not credible; not flexible; vulnerable	Under debate; flexible; under debate

Scholarly debate during the past decade has focused on three aspects of nuclear policy: (1) whether assured destruction could be avoided; (2) whether parity exists between U.S. and Soviet nuclear forces; and (3) whether SALT and related arms control talks are in the best interests of the United States. The debate over these issues is examined in greater detail in Chapter 7. At this point, it should be noted that the past ten to fifteen years have been characterized by intense debate involving incoming political administrations, a wide variety of scientists with backgrounds in weapon systems, a generation of civilian strategists schooled in the theory of nuclear deterrence, and various civilian organizations attempting to influence public opinion and lobby for legislation or changes in policy. As outlined in Chapter 1, at least four major perspectives on nuclear war exist today with widely differing positions on new weapon systems, the significance of new targeting policies, directions in deployment, and so on. While the policies described in this chapter involve substantial debate, the public awareness and level of criticism of the past decade dwarfs that of the earlier period.

Bibliography

The following bibliography includes references cited in the text and related suggested readings; references are organized according to the major divisions of the chapter. The full bibliographic reference for all sources cited in the text of this chapter may be found in either the annotated or unannotated bibliography corresponding to the division of the text in which the source is cited. Sources that are annotated are the most essential or significant.

General Treatments

Freedman, Lawrence. *The Evolution of Nuclear Strategy.* New York: St. Martin's, 1981.

This work traces the varying elements of American nuclear policy between 1945 and 1980. Freedman argues that the initial policies concerning nuclear weapons represented a continuation of a *strategic bombing* orientation that emphasizes offense over defense. Almost all policies concerning nuclear weapons since World War II are presented as problematic given the offensive/defensive character of such weapons. This is a source for understanding early development of nuclear thought, particularly the role of civilian thinkers and the relation between American and European policy concerning use of nuclear weapons.

Head, Richard G. and Erwin J. Rokke, eds. *American Defense Policy,* 3rd ed. Baltimore: Johns Hopkins Univ. Pr., 1965.

Prepared by ''soldiers-scholars'' of the U.S. Air Force Academy, this anthology contains several key documents, from the text of Dulles's initial speech on massive retaliation to that of Melvin Laird on realistic deterrence. In addition to arms control documents (for example, the SALT ABM treaty), this work includes works by politicians, strategists, and academicians: Henry Kissinger, Michael Howard, William Van Cleave, Colin Gray, and Robert Osgood.

Mandelbaum, Michael. *The Nuclear Question: The United States and Nuclear Weapons, 1946–76.* Cambridge: Cambridge Univ. Pr., 1979.

This book examines the role of strategy and diplomacy in dealing with the threat of war. Utilizing a state-centric model, the author argues that the deterrence of war has been strengthened by the existence of nuclear weapons because no way has been found to fight successfully a nuclear war. He views the failure of tactical nuclear weapons, ballistic missile defense systems, civil defense, and counterforce targeting as strengthening the power of deterrence.

Quester, George. *Nuclear Diplomacy: The First Twenty-Five Years.* New York: Dunnellen, 1971.

This book traces the development of the policy of nuclear deterrence. Quester examines the problems of vertical and horizontal arms races, as well as the various attempts at arms control. The author stresses the shaping influence of nuclear weapons on the conduct of international relations.

The Early Cold War Period

Etzold, Thomas H. and John L. Gaddis, eds. *Containment: Documents on American Policy and Strategy, 1945–1950.* New York: Columbia Univ. Pr., 1978.

Indispensable collection of documents (many only recently declassified) from the early Cold War period. Includes essential statements on early U.S. nuclear policies and on U.S.-Soviet relations from the National Security Council (for example, NSC-68), Policy Planning Staff (for example, PPS-7), and Joint Chiefs of Staff (for example, JCS 1952/1). Also includes other influential works from the period, such as Kennan's "The Long Telegram" and "The Sources of Soviet Conduct." Of particular utility are the introductory essays on organizational issues (Etzold) and the strategy of containment (Gaddis).

Kennan, George. *Memoirs: 1925–1950.* Boston: Little, 1967.

Political memoirs by one of America's major figures (as both diplomat and academician) in American-Soviet relations. Includes Kennan's famous communications to Truman regarding what came to be known as containment; also included are essays on Potsdam, the Marshall Plan, NATO, European politics, and more.

Leiberman, Joseph J. *The Scorpion and the Tarantula: The Struggle to Control Atomic Weapons, 1945–1949.* Boston: Houghton, 1970.

Analysis of efforts to control proliferation of nuclear weapons in the immediate postwar period. The role of Niels Bohr, a leading scientist in the production of the atom bomb and an early proponent of creating international structures to control and provide peaceful uses for atomic energy, is described in detail. Major emphasis is also given to the actors and events surrounding the Baruch Plan—the major American attempt at arms control prior to 1950.

Yergin, Daniel. *Shattered Peace: The Origins of the Cold War and the National Security State.* Boston: Houghton, 1977.

This book examines the two major American perspectives toward the Soviet Union and U.S.-Soviet relations during the 1920–50 period. One perspective, viewing the Soviet Union as inevitably hostile and ideologically in opposition to capitalism, is seen as predominating in the period immediately after World War II. A second perspective, emphasizing the legitimate goals of the Soviet Union and implying possible cooperation with the United States, is viewed to have been the prevailing policy assumption in the earlier period.

Alperowitz, Gar. *Atomic Diplomacy: Hiroshima and Potsdam.* New York: Simon & Schuster, 1965.

Baruch, Bernard. "The American Proposal for International Control." *Bulletin of Atomic Scientists* 2 (July 1946): 6.

"The Baruch Plan." *Documents on Disarmament, 1945–1949.* 2 vols. Washington, D.C.: Govt. Print. Off., 1960.

Bernstein, Barton, ed. *The Atomic Bomb: The Critical Issues.* Boston: Little, 1976.

Brodie, Bernard. *The Absolute Weapon.* New York: Harcourt, 1946.

―――――. "Nuclear Weapons: Strategic or Tactical?" *Foreign Affairs* 32 (January 1954): 217–29.

Ferrell, Robert, series ed. *American Secretaries of State and Their Diplomacy.* 20 vols. New York: Cooper Square, 1928–.

Halperin, Morton H. "The Gaither Committee and the Policy Process." *World Politics* 13 (April 1961): 360–84.

Masters, Dexter and Katherine Way, eds. *One World or None.* New York: McGraw-Hill, 1946.

Molotov, V. M. *Speeches at the General Assembly of the United Nations . . . New York, October–December, 1946.* Moscow: Foreign Languages Publishing House, 1948.

Osgood, Robert E. *NATO: The Entangling Alliance.* Chicago: Univ. of Chicago Pr., 1962.

The Pentagon Papers: The Defense Department History of the United States Decision-Making in Vietnam. Senator Gravel edition. Boston: Beacon, 1971.

Spanier, John. *American Foreign Policy Since World War II.* New York: Praeger, 1965.

Sherwin, Michael. *A World Destroyed: The Atom and the Grand Alliance.* New York: Knopf, 1975.

Ulam, Adam. *Dangerous Relations: The Soviet Union in World Politics, 1970–1982.* New York: Oxford Univ. Pr., 1983.

U.S. Department of State. *Foreign Relations of the United States: Diplomatic Papers.* Washington, D.C.: Govt. Print. Off., 1862–.

―――――. *International Control of Atomic Energy and the Prohibition of Atomic Weapons: Recommendations of the United Nations Atomic Energy Commis-*

sion. Department of State Publication 3646. Washington, D.C.: Govt. Print. Off., October 1949.

From Massive Retaliation to MAD

Divine, R. A. *Blowing on the Wind: The Nuclear Test Ban Debate, 1954–1960.* New York: Oxford Univ. Pr., 1978.

An analysis of responses to the testing of the hydrogen bomb, particularly the dilemmas involving radioactive fallout. The author describes the public anxiety over reports of fallout resulting form the atmospheric tests between 1954 and 1958. Major emphasis is placed on the attempts to devise a comprehensive test ban treaty and the reasons only a limited test ban was negotiated.

Enthoven, Alain C. and K. Wayne Smith. *How Much Is Enough? Shaping the Defense Program, 1961–1969.* New York: Harper, 1971.

Written by aides to Defense Secretary Robert McNamara, this book chronicles how policies of MAD (mutually assured destruction) and flexible response developed; generally favorable evaluation of the means and goals of McNamara's strategy.

Kaufmann, William W., ed. *Military Policy and National Security.* Princeton, N.J.: Princeton Univ. Pr., 1956.

Series of essays on American military policy; includes a series of discussions on the importance of developing a "local war" strategy. Most of the articles criticize the idea of massive retaliation, arguing instead for more credible deterrent policies.

_____. *The McNamara Strategy.* New York: Harper, 1964.

An insider's view of early McNamara policies. As one of the leading critics of the Eisenhower/Dulles policy of massive retaliation, Kaufmann strongly shaped McNamara's defense and nuclear policies. This book represents the view of the civilian strategists (for example, particularly from Rand) coming to positions of importance during the early 1960s that rejects massive retaliation and advocates "rational decision-making" in nuclear policy.

Kissinger, Henry A. *Nuclear Weapons and Foreign Policy.* New York: Harper, 1957.

One of the classic texts from the 1950s propounding a variant of limited war theory. Writing for the Council on Foreign Relations, Kissinger argues for limited military and political objectives in foreign policy and seeks to delineate a strategic doctrine in which nuclear weapons have a role in other than all-out wars.

Aron, Raymond. *The Century of Total War.* Garden City, N.Y.: Doubleday, 1954.

_____. *On War.* Translated by Terrence Kilmartin. Garden City, N.Y.: Doubleday, 1959.

Ball, Desmond. *Politics and Force Levels: The Strategic Missile Program of the Kennedy Administration.* Berkeley, Calif.: Univ. of California Pr., 1981.

Brodie, Bernard. *Strategy in the Missile Age.* Princeton, N.J.: Princeton Univ. Pr., 1959.

Gallois, Pierre M. *The Balance of Terror: Strategy for the Nuclear Age.* Boston: Houghton, 1961.

Halperin, Morton H. "Nuclear Weapons and Limited War." *Journal of Conflict Resolution* 5 (June 1961): 146–66.

Kecskemeti, Paul. *Strategic Surrender: The Politics of Victory and Defeat.* Stanford, Calif.: Stanford Univ. Pr., 1958.

Kennedy, John F. *The Strategy of Peace.* New York: Harper, 1960.

King, James E. "Strategic Surrender: The Senate Debate and the Book," *World Politics* 11 (1959): 418–29.

Kissinger, Henry A. "Limited War: Nuclear or Conventional?—A Reappraisal." *Daedalus* 89 (Fall 1960): 800–17.

McNamara, Robert S. *The Essence of Security.* New York: Harper, 1968.

Morgenstern, Oskar. *The Question of National Defense.* New York: Random, 1959.

Schelling, Thomas C. *Nuclear Weapons and Limited War.* Rand Report P-1620. Santa Monica, Calif.: Rand Corp., February 20, 1959.

———— and Morton H. Halperin. *Strategy and Arms Control.* New York: Twentieth Century, 1961.

Schilling, Warner, Paul Hammond, and Glen Snyder. *Strategy, Politics and Defense Budgets.* New York: Columbia Univ. Pr., 1962.

Wasserman, Harvey and Norman Solomon, with Robert Alvarez and Eleanor Walters. *Killing Our Own: The Disaster of America's Experience with Atomic Radiation.* New York: Delacorte, 1982.

Part 2

CONSEQUENCES

CHAPTER 4

The Probability of Nuclear War

S ince the invention of nuclear weapons and the development of nuclear policies, questions have been raised continuously about the likelihood of nuclear war and its controllability if it occurs. It is important to see that the answers to these questions are *estimates* (often quite varied) *of probability*. Once this fact is grasped, one can understand as well that changes in weapons and policies rely on changes in probability estimates. For example, in deterrence theory, one can assume that stability is greatest when the forces on each side are balanced, or that stability is greatest when one side has superior forces. Under these assumptions, those who stress balanced forces see the loss of parity as increasing the likelihood of nuclear war; those who stress superior forces on one side see loss of parity as decreasing the likelihood of nuclear war.

While Chapters 2 and 3 focus separately on the development of nuclear weapons and the development of nuclear policies, this chapter focuses on estimates on (and not responses to) the possibilities for war that arise from successive arrangements of weapons and policies. Of course, nuclear war is a topic around which estimates of probability also take on special import because of concern about consequences (treated in Chapters 5 and 6) and because of the apparent intention of developing weapons and policies that deter war (treated in Chapters 7 and 8).

Despite its close link with these broader issues, the topic of the probability of nuclear war can be treated separately. Perhaps because this topic represents the ''bottom line,'' it receives highest attention among the public and within the popular media. Nevertheless, many people fall into either a *denial* that nuclear war can occur or a *resignation* that nuclear war cannot be avoided. Nuclear planning, however, makes sense only when both of these extremes are rejected. In addition, the *legitimacy* of nuclear weapons and nuclear policies tends to rest on the *probability* that one *assigns* to such issues as outbreak, escalation, and consequences. For this reason, it is important to look initially at how probability estimates are formulated and then at how nuclear war might occur and progress.

This chapter has three sections. First, we provide an overview of probability theory, especially as it comes to merge historically with game theory and nuclear strategy. This initial section provides a context for understanding the use of probability theory outside and within nuclear thought. Second, we survey routes to nuclear war. This survey, which presents systematically the ways nuclear war could begin, cites some of the main popular works on the threat of nuclear war. Concern about this threat is most prominent among critics of nuclear deterrence. Third, we trace scenarios on how nuclear war could be waged. After noting how information on previously secret official scenarios has become available, this section follows the historical development of actual nuclear war plans. As such, this section focuses on the work of those charged with figuring out how to use nuclear weapons in war.

Probability Theory and Nuclear Thought

For many years the paths of probability theory and nuclear thought did not cross. Even today only a small, but important, part of the literature on the nuclear arms race is based on an awareness of probability theory, and this portion of the literature is divided between advocates and critics of applying probability theory to the nuclear arms race. To survey these relations we address (1) the types of probability, and (2) game theory and the rise of formal strategists in the nuclear debate.

Types of Probability

Logicians are careful in using terms like "necessary" (that is, 100 percent probability) and "impossible" (that is, 0 percent probability). Contingent events are ones that are neither necessary nor impossible (their probability is greater than 0 percent and less than 100 percent). Nuclear war is a contingent event; it is neither necessary nor impossible. Hence, either resignation to or denial of nuclear war is illogical. How likely, then, is nuclear war? The problem in calculating the probability of events like nuclear war is how to initially assign the number of "favorable" and possible outcomes. Three theories of probability are generally recognized.[1]

1. Most textbooks on deductive logic include a discussion of types of probability and their limits. For more scholarly approaches the following three works are suggestive of logicians' views. Anton Dumitriv, *History of Logic*, vol. 3, translated by Duiliv Zamfirescu, Dinv Giurcaneanu, and Doina Doneaud (Tunbridge Wells, Kent: Abacus Pr., 1977), pp. 99–101. Dumitriv associates classical probability with Jakob Bernouilli and P. S. Laplace in 1814, frequency probability with R. von Mises in 1928 and subjective probability with J. M. Keynes and B. de Finetti in 1937. A. J. Ayer, *Probability and Evidence* (New York: Columbia Univ. Pr., 1972), p. 27. For the same types of probability, Ayer uses the terms *a priori*, frequency, and judg-

1. Classical theory. The classical theory of probability calculates events in which one can determine the odds without any observation. For example, the probability of getting "heads" in one toss of a coin is 50 percent (one "favorable" and two possible outcomes); the probability of getting a 3 in one roll of a dice is 16.7 percent (one "favorable" and six possible outcomes).

2. Relative frequency theory. The relative frequency theory of probability calculates events in which one can determine the odds on the basis of statistics concerning past occurrences. For example, if a century of recordkeeping shows that a major hurricane hits a coastal state once every decade, then for any given year the probability of this event is 10 percent (one "favorable" and ten possible outcomes).

3. Subjective theory. The subjective theory of probability "calculates" events that lack a solid basis for determining odds. While the calculations make this theory look rigorous and precise, it is pseudoscience. In these events, numbers are hunches or guesses; these estimates are not reliable, and accuracy, if it occurs, is coincidental. Moreover, different people make different assessments, so the numbers are often not consistent. For example, it is mathematically difficult, if not impossible, to calculate reliably that the probability of earth being visited by extraterrestrial intelligence in a given year is .0001 percent.

In the literature on the arms race, few authors indicate what kind of probability they have in mind when they make estimates that they call inevitable, almost certain, unlikely, or impossible. Actually, most of these estimates are hunches—they fall under subjective theory. Readers should generally be cautious of probability estimates; such estimates properly support the argument only if they are reliable.[2] Otherwise, they serve as pseudoevidence. In the next subsection, we address an approach to probability that claims it can be applied constructively to nuclear planning.

Game Theory and Formal Strategy

One area in which subjective probability estimates have been advanced substantially is in game theory. This area is significant because it eventually was incorporated into a highly influential group's approach to nuclear planning. Game theory was developed in the 1920s by the mathematician John von Neumann. Of interest here is the distinction between games of chance and games of strategy. Calculating one's own and oth-

---

ments of credibility. Terrence L. Fine, *Theories of Probability: An Examination of Foundations* (New York: Academic, 1973), especially pp. 230–37. Fine provides a readable survey which is particularly effective in distinguishing and criticizing subjective probability.

2. Estimates of missile accuracy, for example, are reliable. They are based on the relative frequency theory. By test-firing ten missiles, one can draw a circle to measure the distance within which half the missiles land. This circular error probable (CEP) ties the relative frequency of one "favorable" and two possible outcomes, or 50 percent probability, to a measured distance, such as one-quarter mile.

ers' moves in games of chance (for example, matching pennies) is pointless. In games of strategy (such as chess) one's moves need to be made in response to the moves of other players. This part of game theory aims to provide models for calculating the value of various moves in games of conflict.

While some efforts to quantify military planning go back to the early decades of the twentieth century, its major impact was made in the 1950s and 1960s when highly influential civilian strategists tried to apply methods of contemporary social science, particularly economics, to the complexities of nuclear planning, especially to the quantification of deterrence (that is, making mathematical assessments of the advantage and disadvantage of using nuclear weapons in various situations). These social scientific approaches are generally termed systems analysis, and they superseded operational research.[3] Nevertheless, while such methods became widespread in nuclear planning, it was game theory which offered the most mathematically impressive, formal apparatus, although it is debated whether such a model is applicable to nuclear planning.

Game theory had received its classic social scientific application in von Neumann and Morgenstern. In their book, they used game theory to analyze economic problems. As should be obvious, the economy is an arena of conflict, and quantification of strategy seemed promising. However, even in this relatively rational arena, numerous simplifications of the situation and assumptions about human behavior had to be made in order to assign numerical values to the players' options. Despite these problems even in its relation to economic analysis, game theory found its way into military planning. Practitioners of game theory in nuclear policy, particularly Schelling (1960), came to be known as formal strategists because they tried to use formal mathematical models to assess the "nuclear game" in international politics.

In nuclear thought, game theory has been used to claim that the nuclear arms race is rational and nuclear war is improbable. These claims are actually based on comparison of the numbers assigned for the various possible outcomes of the strategies available to the United States and the Soviet Union. "Prisoner's Dilemma" and "Chicken" are two games that are widely used to model strategies of the superpowers. "Prisoner's Dilemma" provides the model for the rationality of the arms race, and "Chicken" provides the model for the improbability of nuclear war.

We do not discuss these games here per se. Rather, we will illustrate how game theory calculates the various outcomes of U.S.-Soviet behav-

---

3. While operational research is based on empirical tests (often war experience), the war games of systems analysis are based on theoretical models. Tests of nuclear weapons have been so small-scale that operational research is rather limited. For a discussion of this point, see Andrew Wilson, *The Bomb and the Computer: Wargaming from Ancient Chinese Mapboard to Atomic Computer* (New York: Delacorte, 1968), especially pp. 45–62.

ior and which strategies it recommends. Basically in games of conflict, one is advised to select the best of the worst outcomes. Often it is recommended that one look at not only how one might win big but also how in so doing one might lose big. For prudence, it is recommended that one accept smaller gains or even some loss to avoid risking a major loss. The technical term of one strategy of this sort is the minimax principle—the rule that one should seek to minimize the maximum loss one's adversary can inflict. While "Prisoner's Dilemma" and "Chicken" are very simple models, superpower relations are quite complex. Moreover, these two games are based on sequential moves (first one side moves, then the other side) and full moves (one side moves to total disarmament and then the other player completes a move in response).

Continuing the arms race is presented as more rational than disarmament because, given the numbers game theorists assign, one risks more by disarming than by arming. Figure 1 illustrates one model and set of numbers that yield this result. Suppose the Soviet Union (a) and the United States (b) each have only two strategies from which to chose: to disarm (1) or to arm (2). Four results are possible: both disarm (a1, b1); only the United States disarms (a2, b1); only the Soviet Union disarms (a1, b2); and neither disarms (a2, b2). Further suppose that the payoff for mutual disarmament is 5 and for mutual armament is minus 5, and that the payoff for nonmutual choices is 10 for the armer and minus 10 for the disarmer. (In each box in figure 1, the upper triangle gives the Soviet payoff, and the lower triangle gives the U.S. payoff.) Under these conditions, while mutual disarmament is preferable, that choice violates limiting the other's potential gain and one's own potential loss—by disarming one will get 5 or minus 10, but by arming one will get minus 5 or 10. According to the minimax principle, one should accept a probable payoff of minus 5 to avoid a possible payoff of minus 10. If one cannot trust that the adversary will disarm, one arms to avoid the worst outcome of disarming while the adversary arms. Of course, one assumes the adversary does likewise. Hence, a mutual arms race is the rational strategy.

The improbability of nuclear war can be concluded assuming a different game. Figure 2 illustrates one model and set of numbers that yield this result. (The structure of the matrix remains constant but the symmetry of the numbers is dropped.) Suppose in a potential nuclear crisis the Soviet Union (a) and the United States (b) each have the options of compromise (1) and steadfastness (2). The four possible outcomes would be: both compromise (a1, b1); only the Soviet Union compromises (a1, b2); only the United States compromises (a2, b1); and neither compromises (a2, b2). Further suppose that the payoff for mutual compromise is 1 (avoid nuclear war) and for mutual steadfastness is minus 100 (wage nuclear war); the payoff for nonmutual choices is 10 for steadfastness and minus 10 for compromise. In this case, the minimax principle would dic-

Figure 1.
Armament vs. Disarmament.

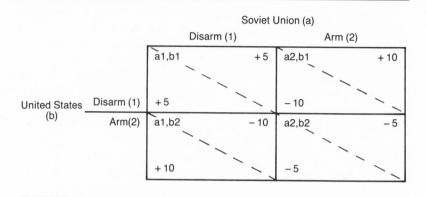

tate that compromise is the rational strategy—it is better to back down and score minus 10 than remain steadfast and risk a score of minus 100. If both sides are rational, mutual compromise will result and nuclear war will be averted.

These two illustrations bring out one of the major problems of game theory, which is assigning numerical value to various payoffs. These values are *subjectively determined.* (Of course, some approaches to non-violent conflict resolution, such as the method often taken in the *Journal of Conflict Resolution,* fall into the same trap—they assign numbers such that nonviolent solutions are more rational.) Moreover, even accepting

Figure 2.
Compromise vs. Steadfastness.

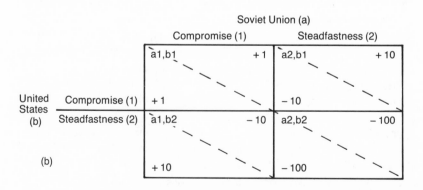

the values as plausible, one must also assume that the other player engages in *rational behavior* and that rational behavior relies on making calculations that are expressed in *numerical terms*—even if this requires hunches.

The approach of game theory, furthermore, stresses *conflict*. While in geopolitics cooperation is found more than conflict, strategists—including those influenced by game theory—tend to lose sight of this fact. Instead, one anticipates the *worst case* and seeks to counter it. Then, one claims the nonoccurrence of the worst case shows *deterrence* works. Generally, the adversary would have little interest in initiating a worst case conflict, but, as Freedman notes:

> The interests that might lead one nation to engage in provocative adventure with a high risk of disaster became less relevant in the analyses than the question of how it might succeed.[4]

This mentality in game theory and strategic thought is closely related to the perspective in political science that is termed *Realpolitik*. This Hobbesian perspective holds that neither moral principles nor international treaties are binding in the geopolitical field. The realist is distrustful, prepares for war, and calculates how to limit the amount of damage that an adversary can inflict. Among game theorists, Rapoport stands virtually alone in seeking to introduce a moral dimension into strategic thought. He sees the role of conscience as a rational factor that can be progressively realized and that undercuts many of the pessimistic strategies that are recommended by game theory's minimax principle.

One final example can illustrate Rapoport's point concerning the limits of game theory. Game theory is only as reliable as the values assigned to various payoffs and the probability of occurrence for each option. For example, on any given day, it will rain or not rain and a person going outside will carry or not carry an umbrella.[5] Suppose one assigns values to each of the four options, as shown in Figure 3. In this figure, taking an umbrella and raining gets minus 2.

How one assigns numbers to values, of course, is quite idiosyncratic. Someone might assign a positive number to taking an umbrella and raining, reasoning that when it rains and you have an umbrella, you get use from it and that is good.

Once values are assigned, calculations can begin. In particular, if one knows the probability of rain for a given day, one can calculate which op-

4. Laurence Freedman, *The Evolution of Nuclear Strategy* (New York: St. Martin's, 1981), p. 180.
5. Here the other player is nature. Moreover, unlike games between the superpowers in which moves are dependent, moves in this game are independent—whether one carries an umbrella does not affect nature's move to rain or not rain. Formally, such games require the Bayesian approach. For our purposes, this game provides an easily understandable way to make our point, and the differences in games that we note here are not relevant.

Figure 3.
Umbrella vs. No Umbrella.

	Rain	No rain
Umbrella	−2	−1
No umbrella	−5	+3

tion is the best: that is, the most rational, with the greatest possible payoff. For example, given the values assigned in figure 3, if the probability of rain is 90 percent, it makes sense to carry an umbrella, but not when the probability is 10 percent.[6] If the probability of rain is uncertain, strategy depends on nonmathematical considerations, for example, moods or morals. If one is a pessimist (as are most game theorists and strategists), one assumes the worst and chooses to carry an umbrella, and one always gets at least a bad result (minus 1) or an even worse result (minus 2). If one is an *optimist*, one does not always choose to carry an umbrella, and one can get a good result (plus 3) or a very bad result (minus 5).

It is wrong to assume that rationality dictates choosing the strategy with the minus 1 or minus 2 payoffs over the strategy with minus 5 or plus 3 payoffs. To make a decision by comparing the two sets of possible outcomes is to abstract from other relevant considerations. Extramathematical considerations are indispensable when the probabilities are not known. In this case, the decisive factor is mood. In nuclear strategy, it would seem a case could be made that morality would provide a more

6. The quantification of this line of reasoning requires that the value under the rain option be multiplied by its probability and added to the value of the no rain option multiplied by its probability. The sum gives the expected utility of that option. One then compares the numbers for taking and not taking the umbrella and chooses the greater payoff. If the probability of rain is 90 percent, then the equation is as follows: with umbrella, $.9 (-2) + .1 (-1) = -1.9$; without umbrella, $.9 (-5) + .1 (3) = -4.2$. If the probability of rain is 10 percent, then the equation is as follows: with umbrella, $.1 (-2) + .9 (-1) = -1.1$; without umbrella, $.1 (-5) + .9 (3) = 2.2$. If the probability of rain is 90 percent, one who takes an umbrella will score -1.9, while one who does not take an umbrella will score -4.2. So it is rational to take the umbrella (-1.9 is greater than -4.2). If the probability of rain is 10 percent, one who takes an umbrella will score -1.1, while one who does not take an umbrella will score 2.2. So, it is rational not to take an umbrella (2.2 is greater than -1.1). This formula is taken from and is further elaborated on by Anatol Rapoport in *Strategy and Conscience* (New York: Schocken, 1964), especially pp. 32 and 33.

appropriate extramathematical component in the decision concerning strategy selection. Nevertheless, it is true but ironic that few people constantly carry umbrellas but most strategists regularly take the pessimistic approach and assume worst case scenarios.

As has been noted, the advocates of applying game theory and other methods of system analysis met with criticism. In particular Rapoport and Green question these methods in detail. An interesting debate on game theory and nuclear strategy has also taken place in philosophy, particularly among utilitarians. One of the first and best critiques of the applicability of game theory per se is by Wolff. An ongoing debate on how to use game theory for or against specific nuclear policy proposals can be found in Lackey (1975, 1982, 1983), Kavka (1980, 1982, 1983), and Hardin.

As the controversy in philosophy indicates, the debate between the defenders and the critics of the applicability of such methods of nuclear planning continues, hinging on whether there is or can be a theory of nuclear deterrence. From an academic perspective, theories that rely on subjective probability estimates that preclude empirical testing are pseudoscientific. If the hypotheses of nuclear strategy are pseudoscientific, it may be deceptive, if not imprudent, to elevate them to the status of theories with the connotation of being subject to testing.

Nevertheless, while this debate over deterrence theory has a forty-year history, many commentators on the nuclear arms race are unaware of it. Since many commentators are rather naive logically and historically in their assertions regarding the likelihood and conduct of nuclear war, one needs to resist the temptation to group commentators merely according to their political positions. A commentator's methodological position is often as indispensable (or even more so) for assessing her or his argument. In the two sections that follow, most of the references to likelihood (how nuclear war could start and how it would be waged) are subjective probability estimates. While most of the routes to and scenarios for nuclear war are possible, they cannot be calculated with precision. Nevertheless, most of the writers considered in these sections use quite specific—often numerical—probability estimates.

What does this mean for the layperson interested in understanding the arms race and the threat of nuclear war? A number of points should be stressed. First, over the past few decades, probability theory, game theory, and nuclear strategy have become highly intertwined. Formal strategies (those that apply elements of game and probability theory to the development of nuclear strategy) have come to play increasingly important roles in the development of nuclear policies, especially force deployment policy. The rise of the formal strategists has had major implications on the idea of deterrence and its role in national security planning; not the least of these implications is that responsibility for security and military planning may have shifted from traditional geopolitical sources to technocrats with little political training.

Second, many of the writings in this area tend to be either highly technical and arcane or oversimplistic, utilizing subjective probability terms giving the appearance (but not the substance) of scientific rationality. An example of this latter pattern may be seen in the destruction projected to be necessary to achieve assured destruction. An arbitrary figure of 60 percent destruction of an adversary's industrial base was once suggested as the minimum level necessary to achieve deterrence. This figure became a basis for which targeting was designed, even though there was no evidence that destruction of 30 or 50 or 80 percent of industrial production would lead to the desired result. Moreover, the technical orientation of such works, even when publicly accessible, means they tend to be read only by those in the strategic planning community.

Third, while there is intense debate among scientists and strategists over the role of probability in nuclear issues, the debate only rarely becomes a matter of public awareness (see Chapter 7). Even when the debate is made public (for example, the current technical responses pro and con to the idea of a strategic defense system), general awareness is often focused narrowly on the proposed "technological fix"—one accepts or rejects the idea that a technological development is not only possible but further that it will solve existing nuclear problems. The overall effect of these developments has been a diminished public understanding. We are all too often left with ignoring the methods and rationales of the strategist or blindly trusting political and military leaders to authoritatively base policies on sound scientific principles.

It is at the same time our purpose here to indicate that the average citizen can become more aware of arguments and able to critically evaluate their implications. Such a critical perspective would minimally include learning the basic types of probability and the methods by which we apply them; analyzing the public literature to ascertain the degree to which nuclear policies are predicated upon subjective estimates; analyzing the degree to which probabilistic models treat or ignore important political and moral dimensions. On the basis of these beginning steps, the average citizen should better be able to address the issues of whether nuclear war could break out and whether it could be controlled, and to comprehend its consequences.

The Threat of Nuclear War: How It Could Start

Since 1947 the *Bulletin of the Atomic Scientists* has kept a "doomsday clock."[7] By moving the hands of the clock closer to or further from

7. A summary of all but one occasion when the hands were moved can be found in *Bulletin of the Atomic Scientists* 36 (January 1980); 2–3.

midnight it assesses the threat of nuclear war. Though not technically a calculation of probability, the closer the hands get to midnight, the less time there is—in the editors' judgment—to prevent war. It is clear that the *Bulletin* regards as threatening the development of more sophisticated weapons and policies of use, while it regards as stabilizing advances in arms control and detente. For example, whereas the hands were set at two minutes to midnight following the development of the hydrogen bomb, they were at twelve minutes to midnight following the signing of the partial test ban treaty. Since 1984, the hands have stood at three minutes to midnight. If and when the hands reach midnight, nuclear war will begin.

Can we know whether or when nuclear war will occur? One answer to this question attracted widespread attention when Schelling et al. (1975) published the article "Nuclear War by 1999?" in which the consensus of the Harvard and MIT professors polled was affirmative. In more recent years the popular literature has often echoed this judgment: see Beres, Calder, Caldicott, Griffiths and Polanyi, and Schell. Many of the following points are treated in varying detail and alarm by these writers.

Calculation of Threat

How great a threat is posed by the fact that nuclear war could begin in many different ways? The human species has always lived under numerous threats. In order to assess how ominous a particular threat is, it must be ranked against other national and global threats. One is the threat of nuclear war, the topic with which this book is concerned. Another is the threat of U.S. or Soviet expansionism. When nuclear war is seen as the primary geopolitical threat, then preventing or rolling back the adversary's expansion is secondary to achieving international arms control. When superpower rivalry or hegemony is seen as the primary geopolitical threat, then international arms control becomes secondary to thwarting the adversary's aggrandizement.

The threat of U.S. or Soviet expansionism generally takes one of three forms. Writers who criticize the United States and sympathize with the Soviet Union often see the greatest global threat as global capitalist colonialism. Writers who criticize the Soviet Union and sympathize with the United States often see the greatest global threat in international communist revolution. A third group often sees the greatest threat in the mutually expansionistic tendencies of the United States and the Soviet Union. The focus here is on the rivalry between the superpowers and its hegemonic consequences for the rest of the world.

Routes to Nuclear War

Concern over the threat of nuclear war usually focuses on how a nuclear war could start. The necessary and perhaps sufficient condition for nuclear war is the detonation of a nuclear weapon in a conflict situation. Three types of detonations of nuclear weapons do not qualify as acts of war. First, a peaceful nuclear detonation (for example, to excavate) is not an act of war. Second, a nuclear test detonation (such as an underground test at a country's own nuclear facilities, in order to learn whether a new warhead design will work) is not an act of war. Third, the accidental detonation of a nuclear weapon on the territory of one's adversary is not an act of war, though it could lead to war. The issue of how initiation could occur concerns, in contradistinction to these cases, the various ways a nuclear detonation could take place in conflict.

The possible combinations of nations in a nuclear war is so large that writers tend to establish types, though most of the types are logically incomplete. However, as treated by most of these writers, there are three types of nations: major nuclear states or the superpowers (the United States and the Soviet Union), minor or secondary nuclear states (China, Great Britain, France, and, perhaps, Israel and India), and nonnuclear states. A nuclear war would involve at least one major or at least one minor nuclear state and could involve one or more nonnuclear states. One suggestive, but not exhaustive, classification of the types of combinations can be found in Beres:

> Two-country nuclear war
> Nuclear attack by a nuclear weapons state against a nonnuclear state
> Nuclear war among several new nuclear powers
> Nuclear war between secondary nuclear powers and one of the superpowers
> Nuclear war between secondary nuclear powers and both of the superpowers
> Nuclear war between secondary nuclear powers plus one of the superpowers and other secondary nuclear powers[8]

Beres, who is perhaps the most rigorous of these writers, also distinguishes how nuclear war could occur at any of these levels for any of the following reasons. First, nuclear war can result from escalation of some lesser conflict. Second, it could result from the accidental use (from mechanical or human malfunction) of nuclear weapons. Third, it could result from the unauthorized use of nuclear weapons. Fourth, it could result from the irrational use of nuclear weapons.[9]

Recent attention has focused on the first two reasons. Since NATO has a first-use policy (it could initiate use of nuclear weapons in response to a conventional attack by Warsaw pact forces), a regional conflict in Europe could result in limited theater nuclear war or could escalate to strategic nuclear war. With regard to accidental nuclear war, mechanical

8. Louis Rene Beres, *Apocalypse: Nuclear Catastrophe in World Politics* (Chicago: Univ of Chicago, Pr., 1980), pp. 157–71.
9. Ibid., pp. 32–73 and 85–95.

malfunction, particularly of computers, receives greatest attention, especially in an era in which the flight time of missiles is often less than the time needed to resolve an erroneous computer indication that nuclear attack is underway.

The third reason largely concerns the effectiveness of C3I systems (communication, command, control, and intelligence). Treatment of the fourth reason is more common in discussions of proliferation and nuclear terrorism than in discussions of nuclear war between the superpowers.

Of course, since nuclear war could occur at any of Beres's six levels and for any of his four reasons, there are at least twenty-four general routes to nuclear war, quite apart from routes involving subnational or terrorist groups with nuclear weapons. These considerations leave out entirely the type of weapons used (tactical or strategic), the type of targets hit (counterforce or countervalue), and the level of kilotonnage or megatonnage exchanged.

Scenarios for Waging Nuclear War

While the preceding section addresses how nuclear war could start, this section concerns how it could progress. Given the very large number of ways nuclear war could begin, there are also a great many ways in which it could proceed. Since early in the history of the nuclear arms race planners have sought to plot the possible progression of nuclear war. In order to understand this topic, it is important to contrast it with three related, but distinct, concerns.

First, scenarios for nuclear war are to be distinguished from theories of deterrence. The latter concern the degree, if any, to which specific nuclear weapons capabilities and policy declarations prevent or discourage nuclear war and prevent or discourage lesser (especially conventional) conflicts. Scenarios for nuclear war concern how strategically one should use nuclear weapons if deterrence fails. For example, a scenario would indicate how the United States could respond to a nuclear attack. Second, scenarios for nuclear war are distinct from criticisms of existing policies and weapons. Because a vast range of options is open, strategic planners seek to determine the most rational course of action at each step along the way. Many times such reflection leads to a call for development of new weapons capabilities or revision of policy. These latter changes in weapons and policies more often reach the level of public debate (at least congressional), whereas the abstract determination and assessment of scenarios tends to occur in less politically engaged military headquarters and think tanks. Third, scenarios are distinct from fantasies. In the context of weapons and policies, scenarios are based on present physical possibilities, not future theoretical possibilities. Such visionary speculation is often provided along with calls for changes in weapons and policies, but is of no immediate benefit in selecting, if necessary, nuclear options.

One final observation is in order. Just as with the section on routes to nuclear war, this one is not focused on consequences. Of course, reflection on scenarios cannot occur apart from knowledge of consequences. Nevertheless, given one's assumptions about the various consequences, a range of options is still open. The next chapter deals with the debate over consequences. Scenarios focus on selection of nuclear options (which weapons against which targets), given existing capabilities. Subsequent policy initiatives such as the current Strategic Defense Initiative (SDI) proposal may well influence Soviet and American planning and lead at some future point to alternative scenarios. But at any one point in time, strategists plan on the basis of existing capabilities of the United States and the perceived capacities of the Soviet Union.

Hiroshima and Nagasaki: The Future of Total War

The first and so far only actual use (as distinct from threatened use) of nuclear weapons occurred in Hiroshima and Nagasaki at the end of World War II. Recent mounting evidence questions the traditional view that the use of these bombs had primarily military purposes, to ensure the defeat of Japan. This recent evidence suggests that the purpose of the bombings was more for political ends, to ensure the total surrender of Japan (as distinct from mere defeat), which until then would have been difficult to achieve on U.S. terms, and secondarily to serve notice to the Soviet Union. The use of these nuclear weapons changed the questions that would be addressed in future reflections on how, if ever, to use nuclear weapons.

In particular, given the devastation of Hiroshima and Nagasaki, the postwar consideration of scenarios for nuclear war focused on whether the use of nuclear weapons would cause total or only partial destruction and on whether war in the nuclear age would be total or could be restricted in its objectives and means. Military and strategic thinking was dominated by the view that modern wars are total.[10] As defined by Osgood, "total war refers to that distinct twentieth-century species of unlimited war in which all the human and material resources of the belligerents are mobilized and employed against the total national life of the enemy."[11] While this view had academic confirmation in Wright and Liddell Hart, and lent support to the official policy of containment (as opposed to rollback of the Soviet Union, particularly by means of use of

10. Bernard Brodie, *Escalation and the Nuclear Option* (Princeton, N.J.: Princeton Univ. Pr., 1966), p. 4.

11. Robert E. Osgood, *Limited War: The Challenge to American Strategy* (Chicago: Univ. of Chicago Pr., 1957), p. 3.

atomic weapons), it was opposed by a perspective that continues to reject the premise that the option facing the United States is one of total war or no war. Included in limited war theories is a wide range of views that share the premise that it is rational to plan to control nuclear use.

After World War II, Brodie (1946) led critics of the dominant view. He argued that current nuclear weapons would not be sufficient to destroy the enemy but that modern war still should be limited. Borden anticipated many subsequent developments in nuclear weapons systems and policies. He gives several quite detailed scenarios and stresses counterforce nuclear attacks (then by aircraft, but later, he speculates, by missiles). Borden's neglected book is perhaps the best point of departure for a look at original sources in the history of writings on scenarios.

Insofar as governmental development of official scenarios is concerned, detailed information is available only after it has been declassified or leaked. As a result more specific and detailed information can be found on the early period than on more recent times. Nevertheless, several researchers have facilitated the task of accessing available information on official scenarios and the data on which they are based. Two groups of researchers deserve special note. On the one hand, several scholars have done much to break down the secrecy surrounding intelligence gathering. In this regard, the works of Bamford, Freedman, and Prados are especially helpful. On the other hand, several scholars have increased greatly our understanding of official scenarios on nuclear war. In this regard, articles by Rosenberg (1979, 1981-82, 1982, 1983) and Mariska, and the book by Pringle and Arkin, which is to date the only book available on the topic, are quite impressive. Much of what follows is based on these works, which were written by persons involved with this planning or who researched recently declassified archive materials on these topics. Most specifically, these writers have made it possible to situate discussion of official scenarios in the context of the single integrated operational plan (SIOP) from Truman through Reagan.

Early Official Scenarios

Shortly after World War II, General Curtis LeMay, as commander of SAC headquarters, was responsible for the preparation of an annual contingency war plan. His scenarios, like the dominant view, assumed total war. Between 1945 and 1949, LeMay developed several scenarios for how to use all of the United States' atomic weapons against countervalue targets (then termed Delta mission) and counterforce targets (then termed Bravo mission) in the Soviet Union.[12]

12. Peter Pringle and William Arkin, *SIOP: The Secret U.S. Plan for Nuclear War* (New York: Norton, 1983), pp. 46 and 56. Much of these details draw on Rosenberg's work, which is more scattered and less accessible.

LeMay worked under the constraints of limited intelligence on the Soviet Union and limited stockpiles of atomic weapons. While maps were old, unreliable, and incomplete, atomic weapons were scarce and nonoperational. For example, between 1945 and 1949 the United States had fewer than fifty atomic weapons, each weapon required two days for assembly, and only thirty B-29 bombers were capable of delivering an atomic bomb. Not surprisingly, in early scenarios the number of planned strikes typically exceeded the number of available atomic weapons. The plans of 1947, 1948, and 1949 illustrate the intent of and problems with early nuclear war scenarios.

In 1947 (when stockpiles were quite low), SAC Emergency War Plan, codenamed "Broiler," planned to hit twenty-four cities and to use thirty-four atomic bombs; in 1948, "Trojan" called for seventy cities to receive one hundred thirty-three atomic bombs; and in 1949, "Off-tackle" targeted one hundred four cities with two hundred twenty bombs, and specified that seventy-two additional atomic bombs be held in reserve.[13] Despite their lack of realism *vis-à-vis* then-current capabilities, LeMay's scenarios became quite detailed from an operational point of view. This detail is illustrated by the now declassified summary of LeMay's 1951 war plan, approved by the Joint Chiefs of Staff on October 22, 1951:

> Heavy bombers flying from Maine would drop 20 bombs in the Moscow-Gorky area and return to the United Kingdom. Simultaneously, medium bombers from Labrador would attack the Leningrad area with 12 weapons and reassemble at British bases. Meanwhile, medium bombers based in the British Isles would approach the USSR along the edge of the Mediterranean Sea and deliver 52 bombs in the industrial regions of the Volga and Donets Basin; they would return through Libya and Egyptian airfields. More medium bombers flying from the Azores would drop 15 weapons in the Caucasus area and then stage through Dhahran, Saudi Arabia. Concurrently, medium bombers from Guam would bring 15 bombs against Vladivostock and Irkutsk.[14]

Just three months before the first Soviet atomic test, explicit public criticism of such scenarios as Trojan and their assumptions surfaced in May 1949 with the Harmon report (see declassified excerpts of this report in Etzold and Gaddis).[15] In addition to its political assessment, the report showed not only that the United States could not execute Trojan but also that even if it could Trojan would destroy "only" 30 to 40 per-

---

13. Ibid., pp. 48–49. "Offtackle" and the earlier "Halfmoon" can be found in Thomas H. Etzold and John L. Gaddis, eds., *Containment: Documents on American Policy and Strategy, 1945–1950* (New York: Columbia Univ. Pr., 1978).

14. Ibid.

15. Freedman, *The Evolution of Nuclear Strategy*, p. 55.

cent of Soviet industry.[16] Scenarios of this complexity and greater complexity (as were becoming the trend, as the above cited plan of 1951 illustrates) would require more weapons and more clearly identified targets.

Still, despite calls for more weapons and targets, between 1945 and 1949 SAC planning and both proponents and critics of the then-dominant view of total war at least shared two assumptions: (1) because of nuclear weapons, the Soviet Union would no longer be the exception to the view that strategic bombardment could destroy any adversary; and (2) because exclusive U.S. monopoly of nuclear weapons for the immediate future seemed evident, neither changes in weapons and policies nor development of more sophisticated scenarios than strategic bombardment with nuclear weapons seemed necessary.[17] It was the latter assumption that was destroyed for all groups in 1949 with the first Soviet atomic detonation.

Impact of Soviet Atomic Test

In 1950, NSC-68 marked the shift from monopoly to stalemate in which, until the late 1960s, the United States enjoyed considerable nuclear superiority. This document, as well as others from the containment era, shows the extent to which U.S. nuclear scenarios were based on misperception and apprehension of Soviet capabilities (they can be found in Etzold and Gaddis). At any rate, NSC-68 spawned more specific scenarios and led to changes in weapons and policies. Because of conventional weakness and nuclear strength, NSC-68 advocated that the United States rely on its nuclear arsenal, and in Europe (NATO), reject a no-first-use position.

The Soviet detonation had also lent support to efforts to develop both the hydrogen bomb and tactical nuclear weapons. It was the latter, once developed, which received important consideration in subsequent scenarios. For example, Project Vista developed a scenario in which use of tactical nuclear weapons would enable the West to preserve Europe from Soviet invasion. This was one of the earliest war games (simulations) that were, for nuclear war, the only substitute for operational experience. In a dozen years the official list of war game models grew to over 200, and recent estimates suggest that in the United States alone "between 15,000 and 30,000 officers and scientists are concerned with war gaming of one kind or another."[18]

The realism of such games, as the Harmon report had made clear, is closely linked not only with the number and types of nuclear weapons but also with the reliability and specificity of data from intelligence gather-

16. Pringle and Arkin, *SIOP*, p. 61. The Harmon report can be found in Etzold and Gaddis, *Containment*.
17. Freedman, *The Evolution of Nuclear Strategy*, p. 93.
18. Wilson, *The Bomb and the Computer*, pp. xii and 59.

ing networks. While the impact on policy and weapons development of Soviet entry into the nuclear club can be traced in the media and congressional reports of the time, the details of the U.S. intelligence effort were and are very closely guarded, although, as noted before, Bamford, Freedman, and Prados have done much in recent years to pull back that veil. At any rate, by the mid-1950s stockpiles of atomic bombs had risen to over one thousand, and several thousand targets, from big cities to small industries and river crossings, had been identified by stepped-up intelligence operations. Among the 1700 prime targets that SAC wanted to be able to hit over several bomber runs were 409 airfields, which shows the increasing shift since 1949 to counterforce targeting.[19]

Truman initiated research on the hydrogen bomb and tactical nuclear weapons and ordered annual refinements of total war scenarios against the Soviet Union, but he was perceived by the public as backing away from reliance on nuclear weapons. As a result Eisenhower became more closely associated in the public eye with total war scenarios. Eisenhower's "new look" assumed that we could not afford to develop both nuclear and conventional forces and had to develop the nuclear option. While criticism of these assumptions was appearing in classified documents, the more important scrutiny of, and debate on, scenarios did not occur until after Dulles's declaration of the doctrine of massive retaliation in 1954.[20] Brodie (1954) was again a leading critic. Before he had supported the theory of limited war. Now he criticized the policy of massive retaliation.

Since the mid-1950s, however, the strategic debate has focused on the size and composition of the defense budget; that is, issues such as the ratio of conventional to nuclear appropriations or the level of interservice rivalry around various weapon systems have tended to dominate political debate.[21] Arguments over whether massive retaliation or limited war policies were more viable are muted or downplayed as assured destruction became the policy. Since that time, with few exceptions, arguments over specific weapons and more recently arms control have occurred, but generally within a policy framework taken for granted. It is within this policy framework that planners of scenarios must work.

SIOP and Limited Nuclear War

During Eisenhower's administration, the number of nuclear weapons grew from about one thousand to about eighteen thousand (targets swelled to twenty thousand) and fierce competition emerged among the army, navy, and air force to develop independent and sufficient targeting scenarios and capabilities. Given the then-staggering overkill that could

19. Pringle and Arkin, *SIOP*, pp. 44 and 56.
20. Brodie, *Escalation and the Nuclear Option*, p. 5.
21. Freedman, *The Evolution of Nuclear Strategy*, p. 93.

result from such overlapping plans, Eisenhower tried to introdce order into attack scenarios by forming a Joint Strategic Target Planning staff.[22] The target list of the group was the basis for the first SIOP. This first SIOP was completed in December 1960 (one month before Eisenhower left office). This and each subsequent SIOP plans communications and command of the army, navy, and air force and control of their nuclear weapons.[23]

While the later Truman administration facilitated massive retaliation, the public associates that policy with the Eisenhower administration. Likewise, the public identifies the incoming Kennedy administration with options other than all or none (limited nuclear war) although it was the outgoing Eisenhower administration that had facilitated a shift to such broadened options. Under Robert McNamara, secretary of defense during the Kennedy and Johnson administrations, SIOP and limited war theory merged.

In rejecting the Eisenhower solution, the Kennedy administration was influenced particularly by Kaufmann (1956). Kaufmann along with Osgood (1954) and Kissinger (1957) had been instrumental in developing the theory of limited nuclear war as an alternative to massive retaliation. While some important distinctions and scenarios can be found in the above works, key to the development of limited war scenarios per se is Wohlstetter's earlier classified study of overseas SAC bases (1954). In his later public essay (1959), he seeks to reclaim retaliation in some forms, since rejection of all war in order to avoid the total war of massive retaliation seemed equally inappropriate. At this time, similar views were also voiced by Kaplan and Schelling. Nevertheless, of greater significance were two books that applied Wohlstetter's approach; Kahn (1969) and Snyder. In comparison, the scenarios in Snyder's work do not come close to Kahn's impressive and influential effort to provide a logically exhaustive typology for possible nuclear wars. Soon, Kahn (1965) had developed scenarios for each of forty-four rungs in a ladder of escalation.

By the summer of 1961, McNamara, as a result of this turn to Rand (the locus for much of the above cited research), had a new version of the first SIOP (which eventually became known as SIOP-62). The options in McNamara's SIOP had five types of targets:

1. Soviet strategic retaliatory forces—missile sites bomber bases, submarine pens, and so on
2. Soviet air defenses away from cities; for example, those covering U.S. bomber routes
3. Soviet air defenses near cities

22. Pringle and Arkin, *SIOP*, pp. 101–4.
23. Ibid., p. 11.

4. Soviet command and control centers
5. If "necessary," all-out "spasm" attack.[24]

Execution of these options was more realistic not only because of growth in stockpiles and targets but also because Eisenhower had gotten the fissionable material into 90 percent of the weapons (whereas previously the two were kept apart, creating a critical lag time in availability of nuclear weapons for use).[25]

Basically, consideration of scenarios since these developments has taken place on two levels: the tactical and the strategic. At each of these levels the critical question is whether control can be maintained once initial use has occurred. During the 1960s, two pro-limited nuclear war books made important arguments that control can be exercised. Knorr and Read provide a detailed consideration of limited strategic war. Brodie (1966) provides an extended argument against the likelihood of escalation and loss of control in tactical nuclear war. In fact Brodie argues that if tactical nuclear weapons fail in their function as a deterrent, they can serve in use as a "deescalating device."[26]

At about the same time, more applied and less theoretical approaches to such game theory were producing contrary results. Military practices of the board type (no troops or weapons) and field type (troops and weapons) are war games in which an exchange is played out (moves and countermoves) until an end is reached. War games tend to be played in such a way that full escalation and heavy destruction occur—neither side backs down. For example, Operation Sage Brush (conducted in Louisiana) and Carte Blanch (conducted in West Germany) resulted respectively in projection of near total fatalities of the state's and country's populations.[27] In other words, the actual psychology of players in war games contradicts the rational decision making imputed to players in the game theory of the formal nuclear strategists. In addition to the critique of the view that nuclear war can be controlled found in such practical military exercises are theoretical critiques of control. In particular, Ball focuses on problems of maintaining C3I (communications, command, control, intelligence) in his theoretical critique of the view that control can be maintained. (Further discussion of C3I can be found in the third section of Chapter 5.)

Following McNamara's SIOP in the Kennedy and Johnson administrations, which stressed counterforce initially (city avoidance) and eventually countervalue (mutual assured destruction), the Nixon and Ford administrations pursued the doctrine of sufficiency, which led to SIOP-5. Often referred to as the "Schlesinger doctrine," the notion of sufficiency was introduced by Kissinger early in the Nixon administration, and, in terms of the

24. Ibid., p. 121.
25. Ibid., p. 110.
26. Brodie, *Escalation and the Nuclear Option*, p. 23.
27. Freedman, *The Evolution of Nuclear Strategy*, p. 109.

SIOP, resulted in increased stress on limited nuclear options and theater war.[28] Sufficiency conceded that there was little prospect for the United States to attain nuclear superiority again, but maintained that it could have enough nuclear weapons for deterrence. On the one hand, the United States had enough nuclear weapons to make the Soviets view its nuclear policies as credible. On the other hand, if deterrence failed, the United States had enough nuclear weapons to execute various nuclear options.

President Carter came into office critical of the emphasis upon strategies for limited nuclear war. He argued instead for a return to minimum deterrence and an increased emphasis on arms control negotiations. This public stance was attacked by those arguing for a new generation of weapons in response to the weapon modernization program of the Soviet Union and an implicit emerging American strategic vulnerability, as well as by those arguing for either continued flexible response or a new strategy to supplant flexible response. Carter's eventual policy position came to be termed a countervailing doctrine: a continuation of the emphasis on limited nuclear war options and counterforce targeting. On the one hand, Carter initiated plans for greater protection for U.S. leadership (Presidential Directive 58) and the general population (Presidential Directive 41). On the other hand, PD-59 shifted targeting away from traditional countervalue targets (economic and population targets) and toward increasing numbers of military and political targets.

Importantly, PD-59 also gave high priority to improvements in command and control efforts and established a planning requirement that American forces be able to survive protracted nuclear war.[29] The presumed intent of these directives was to reduce fatalities if nuclear war occurred. That is, there would be fewer casualties if military and political targets were chosen over population centers and if the Soviet Union also chose to fight with limited military and political targets. This suggested, then, a move away from Carter's earlier interest in deterrence and an increased emphasis on war fighting. It also necessitated new and expanded expenditures for weapons and command and control programs, as seen in the last budget request of his administration.

The Reagan administration came into office critical of Carter's emphasis on arms control, modernization, and overall military and foreign policy. However, the initial policies and directives of this administration indicate a high degree of continuity with Carter's countervailing strategy. While the Carter policies emphasized denying potential Soviet victory, the Reagan administration emphasizes prevailing or winning a protracted nuclear struggle. Rather than parity, the emphasis is now upon superiority. Weapons systems including the MX, Trident II, and Pershing II have been legislated, along with an increased emphasis on improv-

28. Pringle and Arkin, *SIOP,* p. 177.
29. Ibid., p. 186.

ing the command, communications, and control capability of the strategic forces. The more than $20 billion in programs directed toward this C3I primarily focuses on the ability of political and military leaders to survive varying levels of nuclear attack. Thus, the older views of a spasmodic attack have been replaced by an emerging protracted view of nuclear war. Only if one can sustain communication and leadership structures can one project a winning scenario.

In addition, the Carter emphasis on fatality reduction has been extended with new proposals and programs for civil defense (passive defense) and a ballistic missile defense (active defense). Deterrence is presumed to be strengthened by moving away from assured destruction toward more credible options (other than spasmodic attack and retaliation) such as decapitation of Soviet military and political leadership. And should war occur, improved communications coupled with new defense systems could result in survivability and increased chances for winning.

By now the United States has 40,000 potential targets for 10,000 strategic weapons, which range from fifty kilotons to nine megatons and have accuracies between six hundred feet and one mile.[30] Under SIOP-5D, targets include:

1. Soviet nuclear forces: ICBMs and IRBMs, together with their launch facilities and launch command centers, nuclear weapons storage sites, airfields supporting nuclear-capable aircraft, and nuclear missile-firing submarine bases
2. Conventional military forces: caserns, supply depots, marshaling points, conventional airfields, ammunition storage facilities, and tank and vehicle storage yards
3. Military and political leadership: command posts and key communications facilities
4. Economic and industrial targets: (a) war-supporting industries, ammunition factories, tank and armored personnel carrier factories, petroleum refineries, railway yards, and repair facilities; (b) industries that contribute to economic recovery—coal, basic steel, basic aluminum, cement, and electric power.[31]

Actually, urban and political targets can be added but officially are listed as withhold targets for possible negotiation if war is protracted and communication is possible.[32]

Under SIOP-5D the level of basic attack can be at any of four levels:

1. Major attack options
2. Selected attack options

30. Ibid., p. 37.
31. Ibid., p. 187.
32. Ibid., p. 188.

3. Limited attack options
4. Regional nuclear options

This number goes up to six if one adds the level of special attack, which includes:

5. Preemptive attack
6. Launch on warning or launch under attack.[33]

Representative and prioritized targets from each of the four categories are included within these attack options; also included are targets in other Communist countries such as Cuba, China, and the Warsaw Pact countries. Information is not publicly available on the specific number of targets per option delineated in either SIOP-5D or its successor SIOP-6 adopted by the Reagan administration.

A decision about which option to select from each group might well have to be made in a very few minutes, and that decision would have to be communicated and executed in highly adverse conditions. As a result the stress now is more on C3I than on scenarios per se. This point is illustrated by the test conducted early in the Reagan administration. In the biggest command-post exercise in thirty years, "Ivy League 82" was staged in March 1982 (but set in March 1983). The result of this five-day exercise, which included stand-ins for the president and other top aides, was that "commanders had survived long enough in their underground bunkers to fight a drawn-out nuclear war with the Soviet Union."[34] Prevailing in nuclear war has come to mean survival of C3I for sufficient time to execute the SIOP.

Summary

The probability of nuclear war is a topic that receives much attention. Because naive and deceptive use of the language of probability can misrepresent what we know regarding how nuclear war might be initiated and waged, it is important to be aware of the types of probability and the reliability of calculations based on each type. The first section in this chapter showed these types and how issues in the nuclear debate generally involve the imprecise calculations of subjective probability estimates. While the formal strategists did the most to quantify the discussion, even their estimates often relied on subjective criteria that have been widely criticized for their imprecision.

The second section treated the various ways nuclear war could start. We cited logically possible ways and did not assign relative likelihood, though many writers argue that, given current trends, nuclear war before the end of the century is very likely and especially as a result of some

33. Ibid., pp. 187-88.
34. Ibid., p. 22.

computer error or from changes in nuclear weapons, policies, or defenses. The third section surveyed official plans for waging nuclear war. Again, we discussed these plans as physical possibilities and did not comment on the assumptions that such exchanges could be controlled, though the authors of the scenarios we treat tend to view control as likely.

Bibliography

The following bibliography includes references cited in the text and related suggested readings; references are organized according to the major divisions of the chapter. The full bibliographic reference for all sources cited in the text of this chapter may be found in either the annotated or unannotated bibliography corresponding to the division of the text in which the source is cited. Sources that are annotated are the most essential or significant.

Probability Theory and Nuclear Thought

Dresher, Melvin. *Games of Strategy: Theory and Applications.* Englewood Cliffs, N.J.: Prentice-Hall, 1961.

A volume in the Prentice-Hall Applied Mathematics Series, the book's copyright belongs to the Rand Corporation for whom Dresher was a research mathematician; this study was part of a research program for the air force. Interestingly, as Dresher notes, although many of the applications are "discussed in military terms, they can easily be formulated in economic or social science terms" (p. vii). This work is a clear extension of von Neumann and Morgenstern to "military science." This book largely develops the formal side of game theory; it presupposes that the reader has had at least a year of calculus.

Green, Philip. *Deadly Logic: The Theory of Nuclear Deterrence.* Columbus: Ohio State Univ. Pr., 1966.

Classic critique of systems analysis in general as it is applied to nuclear strategy. Gives special attention to Herman Kahn and the social scientific method claimed by him and his followers (less quantitative) and to game theory in nuclear strategy (more quantitative) and its leading figure, Thomas Schelling. Also raises issues in ethical theory and the democratic process to criticize the propriety, in addition to the adequacy, of these various academic supports for the theory of deterrence.

Rapoport, Anatol. *Strategy and Conscience.* New York: Schocken, 1964.

Effort by one of the developers of game theory to show the limits in the application of game theory. A passionate, yet rigorous and detailed, critique of the methods of the formal nuclear strategists. Argues that, when applied to nuclear strategy, the "rational solutions" of game theory (as developed to that point) are too narrow. Introduces extralogical considerations, specifically moral categories. Presents the need to "elevate conscience" into both game theory of strategists and international relations of politicians.

Schelling, Thomas C. *The Strategy of Conflict.* Cambridge: Harvard Univ. Pr., 1960.

By a Harvard economist who was associated also with its Center for International Affairs and who, after a year at Rand, applied game theory to nuclear strategy; this is the classic text in this area. A quite readable, minimally mathematical, argument for a reform of military strategy. Moves from a critique of then-current approaches to international security to a consideration of bargaining and war. Introduces and modifies game theory to handle the problems that are present. Illustrates how his approach works with chapters on topics such as surprise attack and limited nuclear war.

Von Neumann, John and Oskar Morgenstern. *Theory of Games and Economic Behavior.* Rev. Princeton, N.J.: Princeton Univ. Pr., 1953.

Classic application of game theory to a social scientific arena (economic behavior). This highly formal development of game theory provided the terms and models for studies by others of political and military conflict. Especially relevant, though mathematically complex, is Chapter 11, "General Non-Zero-Sum Games," in particular Section 60, "The Solutions of All General Games with $n < $ or $= 3$" and specifically 60.2, "The Case $n = 2$" and 60.4, "Comparison with the Zero-Sum Games." (For strategy with nuclear weapons, $n = 2$ and, given consequences, game is non-zero sum.)

Ayer, A. J. *Probability and Evidence.* New York: Columbia Univ. Pr., 1972.

Dumitriv, Anton. *History of Logic.* Translated by Duiliv Zamfirescu, Dino Giurcaneanu, and Doina Doneaud. Tunbridge Wells, Kent: Abacus Pr., 1977.

Fine, Terrence L. *Theories of Probability: An Examination of Foundations.* New York: Academic, 1973.

Hardin, Russell. "Unilateral versus Mutual Disarmament." *Philosophy and Public Affairs* 12 (1983): 236–54.

Kavka, Gregory S. "Deterrence, Utility, and Rational Choice." *Theory and Decision* (1980): 41–60.

_____. "Doubts about Unilateral Nuclear Disarmament." *Philosophy and Public Affairs* 12 (1983): 255–60.

_____. "Some Paradoxes of Deterrence." *The Journal of Philosophy* 75 (1978): 285–302.

Lackey, Douglas P. "Disarmament Revisited: A Reply to Kavka and Hardin." *Philosophy and Public Affairs* 12 (1983): 261–65.

_____. "Ethics and Nuclear Deterrence." In *Moral Problems.* Edited by James Rachels. New York: Harper, 1975.

_____. "Missiles and Morals: A Utilitarian Look at Nuclear Deterrence." *Philosophy and Public Affairs* 11 (1982): 189–231.

Maistrov, L. E. *Probability Theory: A Historical Sketch.* Translated and edited by Samuel Katz. New York: Academic, 1974.

Quade, E. S. and W. I. Boucher, eds. *Systems Analysis and Policy Planning: Applications in Defense.* New York: American Elsevier Pub. Co., 1968.

Wolff, Robert Paul. "Maximization of Expected Utility as a Criterion of Rationality in Military Strategy and Foreign Policy." *Social Theory and Practice* 1 (1970): 99–111.

The Threat of Nuclear War

Beres, Louis Rene. *Apocalypse: Nuclear Catastrophe in World Politics.* Chicago: Univ. of Chicago Pr., 1980.

Well-researched and documented study by one of the most prolific and erudite critics of nuclear weapons. Traces three routes to nuclear war (between superpowers, through proliferation, and by terrorism). Details the consequences of nuclear war as a result of each of these routes. Proposes ways to prevent nuclear war at each of these levels.

Calder, Nigel. *Nuclear Nightmares: An Investigation into Possible Wars.* New York: Viking, 1979.

This book by a popular science writer is a competent, easily read survey of the more commonly cited routes to nuclear war (for example, the European theater).

Caldicott, Helen. *Nuclear Madness! What You Can Do!* New York: Bantam, 1980.

Passionate manifesto by the founder of Physicians for Social Responsibility. Brief, readable, alarmist, and at times exaggerated, yet for many readers an illuminating and inspirational introduction to antinuclear activism.

Schell, Jonathan. *The Fate of the Earth.* New York: Knopf, 1982.

Highly influential, passionately written, frequently insightful—as well as repetitious—assessment of nuclear war and global alternatives. Widely praised for raising important questions but controversial because of its very pessimistic postattack ecological assumptions and its very insistent rejection of the traditional concept of national sovereignty.

Schelling, Thomas, et al. "Nuclear War by 1999? *Harvard Magazine* (November 1975): 19-25.

Lively and informative discussion on the nuclear threat by leading professors at Harvard and MIT who have been involved in nuclear planning. Nearly all see nuclear war as likely before the twenty-first century.

Etzold, Thomas H. and J. H. L. Gaddis. *Containment Documents on American Policy and Strategy, 1945-1950.* New York: Columbia Univ. Pr., 1978.

Gerson, Joseph. *The Deadly Connection: Nuclear War and U.S. Intervention.* Cambridge, Mass.: American Friends Service Committee, 1983.

Griffiths, Franklyn and John C. Polanyi, eds. *The Dangers of Nuclear War.* Toronto: Univ. of Tornto Pr., 1979.

Russett, Bruce. *The Prisoners of Insecurity: Nuclear Deterrence, the Arms Race, and Arms Control.* San Francisco: Freeman, 1983.

Thompson, E. P. and Dan Smith, eds. *Protest and Survive.* New York: Monthly Review Pr., 1981.

Scenarios for Waging Nuclear War

Borden, William L. *There Will Be No Time.* New York: Macmillan, 1946

One of the earliest works on nuclear strategy, this, until recently, largely neglected book was influenced by Brodie's *The Absolute Weapon* (below). Borden's counterforce scenarios and speculations on development and use of missiles demonstrate the degree to which he anticipated much of the subsequent development of nuclear weapons systems and policies.

Brodie, Bernard, ed. *The Absolute Weapon: Atomic Power and World Order.* New York: Harcourt, 1946.

This early and very influential book on nuclear strategy is closely linked with the perspective of Brodie, the book's editor and main contributor. It also includes pieces by F. Dunn, A. Wolfers, P. Corbett, and W. Fox. The book argues that nuclear weapons (then only the atomic bomb delivered by aircraft) are not absolute in the sense of preventing or winning war. The book argues for limited use (largely counterforce to avoid destruction of population). Among other things, Brodie cites and details eight ways nuclear weapons have changed war.

————. "Nuclear Weapons: Strategic or Tactical?" *Foreign Affairs* 32 (January 1954): 217–29.

Following his internal documents at Rand since January 1952 (which were classified because thermonuclear weapons were not tested until November 1952 and not publicly announced until 1954), this is Brodie's first openly published article on his criticism of the official views. Argues for utility of tactical nuclear weapons and for pursuit of options for limited nuclear war.

Halperin, Morton H. *Limited War in the Nuclear Age.* New York: Wiley, 1953.

This book is useful for two specific reasons. First, it provides an overview of the discussion of limited war. Though its focus is broader than nuclear strategy, it contains an important treatment of nuclear weapons and local war. Second, it provides a nearly complete annotated bibliography of the public writing to September 1962 on limited war.

Kahn, Herman. *On Thermonuclear War.* Princeton, N.J.: Princeton Univ. Pr., 1960.

One of the most frequently cited and influential books on nuclear strategy, especially for its reflections on scenarios. It and Snyder's *Deterrence and Defense* (below) are the two primary endeavors to apply Wohlstetter's argument (below) that scenarios for less than all-out nuclear war need to be planned. Kahn's contribution is generally judged to be superior because of his aim to be logically exhaustive.

Knorr, Klaus and Thornton Read, eds. *Limited Strategic War.* New York: Praeger, 1962.

This collection addresses scenarios for waging nuclear war that use strategic (as opposed to tactical) weapons and that are limited (as opposed to total). Basically, the prospect for limited strategic counterforce exchanges is analyzed.

Mariska, Mark. "The Single Integrated Operational Plan." *Military Review* (March 1972): 32–39.

This brief article is written by a member of the Joint Strategic Target Planning Staff at Offutt Air Force Base, which is charged with developing targets and attack plans for all U.S. nuclear forces. It gives an authoritative look at the divisions in and work of those who develop the SIOP.

Pringle, Peter and William Arkin. *SIOP: The Secret U.S. Plan for Nuclear War.* New York: Norton, 1983.

The first comprehensive examination and critique of the history of SIOP from the Truman through the Reagan administrations. Shows how the number of weapons and targets have grown and why and how a single plan with various options was developed. Traces the role of intelligence gathering in nuclear strategy. Details crucial status of C3I (communication, command, control, and intelligence) to nuclear strategy. Concludes nuclear war likely will become full-scale, but C3I likely will remain sufficiently intact long enough to execute the SIOP options selected.

Rosenberg, David Alan. "The Origins of Overkill: Nuclear Weapons and American Security 1945–1960." *International Security* 7 (Spring 1983): 3–71.

This virtual monograph brings together for the first time a detailed summary of recently declassified yet still largely inaccessible documents on official U.S. nuclear scenarios from 1945 to 1960. An example of meticulous research into military and nuclear strategy with detailed references to the archival documents.

Snyder, Glenn. *Deterrence and Defense: Toward a Theory of National Security.* Princeton, N.J.: Princeton Univ. Pr., 1961.

This book and Kahn's *On Thermonuclear War* (above) are the two primary endeavors to apply Wohlstetter's (below) argument that scenarios for less than all-out nuclear war need to be planned. Though it distinguishes deterrence by denial and deterrence by punishment, this book is much less detailed than Kahn on scenarios.

Wilson, Andrew. *The Bomb and the Computer: Wargaming from Ancient Chinese Mapboard to Atomic Computer.* New York: Delacorte, 1968.

A general historical overview of war games with an emphasis on nuclear war games, especially as aided by computers. Notes how nuclear weapons are not well suited to operational research and how systems analysis took over in planning strategy. Explores the limits of war gaming with computers.

Wohlstetter, Albert. "The Delicate Balance of Terror." *Foreign Affairs* 37 (January 1959): 211–34.

Public version of his argument that limited, not massive, retaliations needed to be planned. Influenced the development of scenarios such as those in Kahn and Snyder (above).

————, et al. *Selection and Use of Strategic Air Bases.* Rand Report R-266. Santa Monica, Calif.: Rand Corp., April 1954.

Originally a classified study of SAC's overseas air bases, analyzing how to select optimal locations for strategic forces. One of the sources out of which deterrence theory developed.

Bamford, James. *The Puzzle Palace: A Report on NSA, America's Most Secret Agency.* Boston: Houghton, 1982.

Brodie, Bernard. "Strategic Bombing: What It Can Do." *The Reporter* (August 15, 1950).

————. "Strategy Hits a Dead End." *Harper's Magazine* (October 1955): 33–37.

————. "Unlimited Weapons and Limited Wars." *The Reporter* (November 18, 1954): 1–28.

Fialka, Jon J. "Nuclear Reaction: U.S. Tests Response to an Atomic Attack." *Wall Street Journal* (March 26, 1982).

Freedman, Lawrence. *U.S. Intelligence and the Soviet Strategic Threat.* London: Macmillan, 1977.

Kahn, Herman. *Thinking about the Unthinkable.* New York: Horizon, 1962.

Kaufmann, William, ed. *Military Policy and National Security.* Princeton, N.J.: Princeton Univ. Pr., 1956.

Kissinger, Henry. *Nuclear Weapons and Foreign Policy.* New York: Harper, 1960.

Knorr, Klaus, ed. *NATO and American Security.* Princeton, N.J.: Princeton Univ. Pr., 1959.

LeMay, Curtis with MacKinley Kantor. *Mission with LeMay: My Story.* Garden City, N.Y.: Doubleday, 1965.

Liddell Hart, B. H. *Strategy.* New York: Praeger, 1946.

Osgood, Robert E. *Limited War: The Challenge to American Strategy.* Chicago: Univ. of Chicago Pr., 1957.

Prados, John. *The Soviet Estimate: U.S. Intelligence Analysis and Russian Military Strength.* New York: Dial, 1982.

Rosenberg, David Alan. "American Atomic Strategy and the Hydrogen Bomb Decision." *The Journal of American History* 66 (June 1979).

_____. "A Smoking Radiating Ruin at the End of Two Hours: Documents on American War Plans for Nuclear War with the Soviet Union 1954–55." *International Security* 6 (Winter 1981 to 1982): 3–17.

_____. "U.S. Nuclear Stockpile, 1945 to 1950." *Bulletin of the Atomic Scientists* (May 1982): 25–30.

Schelling, Thomas C. *Nuclear Weapons and Limited War.* Rand Paper P-1620. Santa Monica, Calif.: Rand Corp., February 20, 1959.

CHAPTER 5

Consequences of Nuclear War

In most minds questions regarding the probability of nuclear war are closely associated with questions concerning the consequences of nuclear war. The topic of consequences has received far more attention, largely because more empirical data and theoretical models are available for specifying actual and possible consequences. The topic of consequences, however, can be divided into the direct postattack effects of nuclear war and into the indirect preattack effects of the nuclear arms race itself. This chapter addresses the more typical concern over the postattack effects. Chapter 6 covers the preattack effects of the nuclear arms race.

Our knowledge of the consequences of a nuclear war is based on three sources: (1) studies of actual nuclear weapons tests (including the bombings of Hiroshima and Nagasaki); (2) studies of limited and full-scale nuclear war, which normally include projections and extrapolations of casualties, damage to military and industrial bases, and so on, and necessarily involve estimates of probable (likely) consequences; and (3) computer simulations and speculative models of possible catastrophic consequences (for example ozone depletion and nuclear winter). Figure 4 illustrates many of the short-term and long-term effects of nuclear weapons. Most studies of these effects include low, medium, and high alternative models; the effects vary as a function of factors such as the number and size of weapons detonated, climatic conditions, density of population targeted, and atmospheric or ground detonation. It is important to note that in the vast majority of studies, even low-level nuclear exchange (particularly when strategic weapons were used) would result not only in a level of destruction that would equal, if not dwarf, previous conflicts in as little as a few minutes or hours, but would also be characterized by effects not normally experienced in conventional wars: radiation, genetic damage, ecological destruction, epidemics and the collapse of medical care, predominantly civilian casualties, and potentially catastrophic environmental changes.

Figure 4.
Nine Barriers to Well-Being.

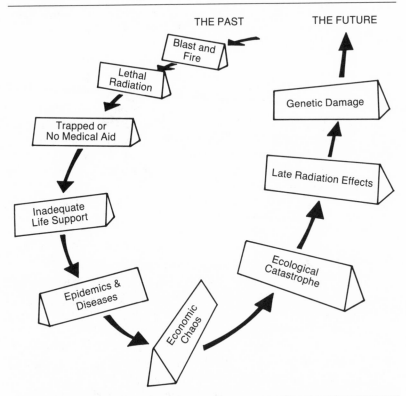

THE PAST   THE FUTURE

Blast and Fire

Lethal Radiation

Genetic Damage

Trapped or No Medical Aid

Late Radiation Effects

Inadequate Life Support

Ecological Catastrophe

Epidemics & Diseases

Economic Chaos

Source: Department of Defense, *DCPA Environment Attack Manual* (Washington, D.C.: Govt. Print. Off., 1973), Chapter 8, Panel 1.

Each of the effects illustrated in Figure 4 as barriers is applicable whether using a low, medium, or high model of probable damage. And while the magnitude of effects multiply as one moves toward medium and high scenarios, it remains true that the consequences represent a fundamentally different level of destruction relative to most conventional wars. Unintentionally, the development of these hierarchial scenarios can leave the reader with the mistaken sense that "limited" use of nuclear weapons would result in damage not significantly different than previous wars. Estimates of casualties that range from 2 to 50 percent of the population obscure the fact that one bomb of average size detonated on a large American urban area such as Chicago would result in more casualties than the combined casualties of all previous wars in which this country has been involved during the past two centuries. One submarine today can do more damage than the total American nuclear arsenal that existed in the early 1950s. In contrast, some writings on consequences imply that a nuclear war would necessarily threaten the human species or the planet itself. Most of the evidence suggests this is unlikely; however, recognition that there would be survivors does not mean that nuclear war is justified on moral, political, or national security grounds, nor winnable in the traditional use of the concept.

In discussing the effects and consequences of nuclear war, it is important to keep in mind three points. First, a substantial amount of knowledge about the consequences of nuclear war is conjectural—what we categorize below as probable and possible. Second, what we do know empirically from tests and studies of Hiroshima and Nagasaki indicates that a nuclear war is fundamentally distinct in character, particularly in its effects on survivors and the environment. Finally, the magnitude of the effects are such that a comparison of low, medium, or high scenarios of damage is of less significance than the understanding that even limited nuclear war has consequences significantly more severe than the worst of most conventional wars.

Nuclear Weapons Tests

While quite a few studies have been made that deal with actual nuclear weapons tests, their publication or declassification often postdates by many years the tests they analyze. The reason is twofold. On the one hand, much information on tests is initially classified for reasons of national security. As this information is declassified, we can get a detailed look at the data on tests, especially U.S. aboveground tests between 1945 and 1962. On the other hand, the effects of nuclear tests continue to exert themselves for many years. As a result, adequate analysis can span many decades. At present, no studies cover more than a forty-year period, but the effects of the atomic bombing of Hiroshima and Nagasaki will continue even beyond the death of survivors because genetic damage extends to future generations with as yet unknown consequences.

Changes in the technology of nuclear devices make assessment of consequences more problematic. For example, the first major study, *The Effects of Atomic Weapons* (Glasstone et al.), was published in 1950 and considered at that time to be the definitive text on the effects of atomic or fission weapons. It was outdated by 1952 when the first hydrogen or fusion weapon was tested. From 1957 to the present, *The Effects of Nuclear Weapons* by Glasstone and Dolan has been the most authoritative technical source.

Physical Effects

Discussion of the physical effects of any nuclear weapon can generally be treated under one of four topics: blast and shock, thermal radiation, initial nuclear radiation, and residual nuclear radiation.[1] Blast and heat account for about 85 percent of the energy. (More technically, blast and shock account for 50 percent, and thermal radiation accounts for 35 percent.) The other 15 percent is in the form of initial (5 percent) and residual (10 percent) radiation. The levels of each of these effects depend primarily upon the power of the blast (which is expressed in terms of its equivalence to TNT and ranges from kilotons to megatons) and the type of blast (air, surface, or subsurface). Generally, these effects can be expressed in human terms by showing the expected percentages of immediate fatalities as a function of distance from the detonation. For example, a one-megaton surface blast on a city would yield the following: (1) a crater .24 miles in diameter, with a fireball reaching .07 miles in radius; (2) out to 1.7 miles, 98 percent dead and 2 percent hurt; (3) between 1.7 and 3 miles, 50 percent dead and 40 percent hurt; (4) between 3 and 5 miles, 5 percent dead and 45 percent hurt, (5) between 5 and 7 miles, 25 percent hurt; (6) out to 5 miles there would be many fires that potentially could spread out to 7 miles.[2] If a firestorm occurs, the above area of lethality would increase fivefold.[3]

To these immediate effects must be added the effects of radiation, particularly fallout. Levels of radiation depend on the weight and type of blast and also on wind speed and direction. Generally, the pattern is cigar-shaped, with one end at the point of detonation and the other at some point downwind, and effects on persons are measured in terms of level of radiation exposure as a function of time. For example, the following radiation levels would be expected from a five-megaton surface blast with a fifteen-mile per hour wind: (1) peak radiation of 500 roentgens per hour (R/hr) would be expected in four hours at points forty miles downwind of the blast; (2) fifty R/hr in eight hours at 100 miles, (3) five R/hr in twelve hours at 160 miles, and (4) .5 R/hr in sixteen hours at 220 miles.[4] An exposure to 600 to 1000 roentgens of radiation in a short period of time results in death within two months for 80 percent to 100 percent of those exposed. Downwind from a blast, fallout will significantly increase the level of early fatalities among the unprotected population.

In the example given here, the fallout pattern would extend in less than a day to an area about 300 miles in length and 100 miles in width (30,000 square miles). Beyond this area, many more persons who received much lower levels of exposure to radiation will die over the next

1. An additional physical effect is electromagnetic pulse, which is discussed in the final section of this chapter.
2. U.S. Department of Defense, Defense Preparedness Agency, *DCPA Attack Environment Manual* (Washington, D.C.: Govt. Print. Off., 1973), Chapter I, Panel 8.
3. Ruth Adams and Susan Cullen, eds., *The Final Epidemic: Physicians and Scientists on Nuclear War* (Chicago: Educational Foundation for Nuclear Science, 1981), p. 175.
4. U.S. Department of Defense, *DCPA Attack Environment Manual*, Chapter 6, Panel 11.

several decades from various forms of cancer that result from the fallout. Studies on postwar nuclear tests in the Pacific and Atlantic oceans and in the United States provide further understanding of physical effects. For example, Berkhouse et al. treat the first postwar nuclear weapons test series, and Gladeck et al. treat the first test of a hydrogen bomb. Some noteworthy related studies include Conrad et al., Hines, and Schultz.

Sociomedical Effects

The atomic bombings of Hiroshima and Nagasaki had measurable consequences that offer the most extensive data to date on the sociomedical effects of nuclear blasts on human communities. While there is a great deal of available information, it is rather gruesome and difficult to summarize. Nevertheless, there are many poignant overviews of the types of human problems that had to be faced. For example, Douglas Lackey gives a brief account of and feel for the sociomedical, as well as moral, issues precipitated by the bombing of Hiroshima:

> When the first atomic bomb exploded over Hiroshima on the sixth of August, 1945, the temperature at the point of detonation reached 50 million degrees. It was 8:15 in the morning; people were on their way to work, and children were on their way to school. The explosion released a blinding flash of light and a searing wave of heat. The concussion that followed leveled a square mile of buildings in the center of the city; the firestorm that ensued burned for six hours and consumed 3.8 square miles of houses and offices. The radiation generated by the bomb was neither seen nor heard, but it was fatal at distances of half a mile.
>
> 80,000 Japanese were killed at Hiroshima, and 100,000 more were injured. Among the dead and injured were tens of thousands of children. The children are especially worth thinking about, since only a perverse logic could blame *them* for the attack on Pearl Harbor. Thousands of children died immediately from the blast or from falling debris. Thousands more suffered flesh burns and died after several hours, or even days, of unrelievable pain. Thousands more were overcome by radiation sickness and died after days or weeks of dehydration and vomiting. Tens of thousands more were blinded, maimed, or were permanently disfigured by keloid scars caused by burns inflicted by the bomb which the Americans had called *Little Boy*.[5]

On the basis of such consequences in Hiroshima and similar ones in Nagasaki, physicians have been able to calculate the range of medical problems that virtually any community would face after even a single nuclear blast. Jack Geiger offers the following summary:

> In addition to third-degree burns, . . . ''survivors'' would suffer crushing injuries, simple and compound fractures, penetrating wounds of skull, thorax and abdomen, and multiple lacerations with extensive hemmorhage, primarily

5. Douglas Lackey, ''The Moral Case for Unilateral Nuclear Disarmament,'' *Philosophy and Social Criticism* 10, Nos. 3 and 4 (1984): 157–58.

in consequence of blast pressures and the collapse of buildings. . . . A moderate number would have ruptured internal organs, particularly the lungs, from blast pressures. Significant numbers would be deaf in consequence of ruptured eardrums, in addition to their other injuries, and many would be blind since—as far as 35 miles from ground zero—reflex glance at the fireball would produce serious retinal burns.

Superimposed on these problems would be . . . cases of acute radiation injury, superficial burns produced by beta and low-energy gamma rays, and damage to radionucleides in specific organs. Many would die even if the most sophisticated and heroic therapy were available; others, with similar symptoms but less actual exposure, could be saved by skilled and complex treatment. In practical terms, however, there will be no way to distinguish the lethally-irradiated from the non–lethally-irradiated.

Finally, this burden of trauma will occur in addition to all pre-existing disease among "survivors," and this list of problems is not based on consideration of the special problems of high-risk populations—the very young and the very old, for example—which are particularly vulnerable.

These are the short-range problems to which a medical response must be addressed. But who will be left to respond?

Physicians' offices and hospitals tend to be concentrated in central-city areas closest to ground zero. If anything, physicians will be killed and seriously injured at rates greater than those of the general population, and hospitals similarly have greater probabilities of destruction or severe damage. . . .[6]

Such severe medical crises lead almost immediately to acute social problems. To a large extent, the social consequence of nuclear attacks on populations is the breakdown of those communities. In most cities, a single nuclear weapon can destroy or damage most buildings, including medical facilities. With so many people dead or injured and so many facilities destroyed or damaged, many families and most social organizations will collapse, at least as functional units. Beyond such medical and social crises are the economic obstacles that will have to be faced by each surviving remnant of the former communities. The destruction of communities also entails the loss of most of their economic assets. For years survivors will face a precarious economic future which will further increase their already high level of psychological stress.

One of the most dramatic and detailed studies of physical, medical, and social effects is that prepared by the Committee for the Compilation of Material on Damage caused by the Atomic Bombs in Hiroshima and Nagasaki. Whereas Glasstone and Dolan focus primarily on the physical effects of nuclear weapons in general, this committee deals largely with the medical and social effects on the people who were in Hiroshima and Nagasaki at the time of the bombings or who arrived shortly thereafter. *Hiroshima and Nagasaki*, the book published by the committee, super-

6. Jack Geiger, "Illusion of Survival" in Adams and Cullen, *The Final Epidemic*, pp.176–77.

sedes the previous voluminous literature on the effects of these bombings. The Japan National Preparatory Committee provides earlier and
less detailed documents on the physical, medical, and social effects of
the atomic bombings of Hiroshima and Nagasaki but is distinctive because of its more explicit political character and advocacy of an international movement to ban all nuclear weapons.

Psychological Effects

Survivors of the use of nuclear weapons and populations threatened
by the existence of nuclear weapons show the adverse psychological effects of nuclear weapons. Clearly the more pervasive psychological effects are presently being caused by the superpowers' stockpiling of nuclear weapons. Persons living in the nuclear age face the possibility of
the destruction of their world—the past and future of their culture can be
annihilated in a blinding, burning, irradiating flash. For forty years, part
of growing up has involved becoming aware of the precariousness of all
that we love. When faced, this awareness takes away the sense of safety
and security that homes, families, routines, communities, and cultures
provide. Psychologically, the threat of nuclear war has brought about,
on a global scale, a form of collective neurosis.

On both the individual and cultural level, escapes from this anxiety
are pursued. The common defense mechanism is termed psychic numbing. More precisely, psychic numbing is the result of repressing emo-
tional reaction to and intellectual reflection on the nuclear threat. Psychologically, one is able to block or reduce awareness of the threat—one
avoids focusing on the horror of nuclear war and its aftermath. Nevertheless, psychic numbing takes its toll. Many children have an ambigu-
ous sense of the future, many young people try to avoid facing the incon-
sistency in their beliefs that they too can pursue a profession and that
nuclear war will occur in their lifetime, and many adults feel frustrated
that they cannot promise their children a world with a future.

Of course, the use of nuclear weapons causes much more severe psychological problems than those that result from the stockpiling of nuclear
weapons. Studies of the psychological effects of the atomic bombings of
Hiroshima and Nagasaki reveal on a small scale the tremendous mental
strain of living that faces survivors of a nuclear attack. Most survivors
are plagued by death imagery because they saw so much death and destruction. They may even fail to clearly distinguish life from death and
sense death in life as their reality. To stay sane, many will turn to strong
forms of psychic closing off. While this effort cuts off the death imagery,
it leaves people going about without feeling. Eventually, as survivors
begin to reflect and seek to regain some semblance of normalcy, they
face feelings of guilt and shame that they survived and are recovering.
There will be a need to engage in a futile effort to justify one's own sur-

vival. For many, these psychological strains will be too great. For others, their injuries and losses will in themselves precipitate psychological crises. Only a few will receive the comfort and support they need, let alone psychiatric care. While the survivors of Hiroshima and Nagasaki could receive help from the outside and could eventually lose themselves in another place, survivors of a major nuclear war would receive little, if any, outside help, and for many there would be no place to go to escape the external reminders of the psychological burdens they bear.

Lifton did the pioneering study on the psychological effects of nuclear attacks. His terminology, such as psychic numbing in reference to both preattack and postattack effects, has become standard. *Hiroshima and Nagasaki* provides further details. More recent studies of the effects of the nuclear threat on populations, including children, can be found in Chivian and Goodman.

Probable Consequences of Nuclear War

On the basis of data gathered from studies of nuclear weapons tests it has been possible to project the consequences of multiple detonations and to speculate on the likely results of various types of nuclear war. Without exception researchers conclude that whether limited or full-scale, nuclear war will bring about disastrous consequences unprecedented in the history of warfare. Many more people will be killed in much less time than ever before—hundreds of millions of people could die in a nuclear war that lasts only a few days. The superpowers, as functional political entities, could be destroyed. Fallout will deposit lethal radiation in nations far from the conflicts. The ecosystem will be disturbed in ways that will render some species extinct and severely tax many others.

Limited Nuclear War

Although attention is often focused on the results of single nuclear detonations and the consequences of large-scale nuclear war, some efforts have been made to project the effects of limited nuclear war. Typically, government estimates of civilian casualties and industrial destruction are much more optimistic than those provided by the scientific community, which also tends to question whether scenarios for limited nuclear exchanges are even credible. For example, in 1974 then-secretary of defense James Schlesinger provided the following calculations. He postulated that the Soviets might target one one-megaton warhead on each U.S. ICBM silo. For five of the six Minuteman fields, he projected only 300,000 fatalities and 500,000 casualties. Adding in estimates for the sixth, which is closest to a major population center, Schlesinger placed total U.S. fatalities at 800,000 and casualties at

1,500,000.[7] His calculations for one one-megaton warhead against each SAC bomber base yielded only 300,000 fatalities and 700,000 total casualties.[8] When a panel of scientists in the civilian sector reviewed this report, they concluded that "the casualties calculated were substantially too low for the attacks in question as a result of a lack of attention to intermediate and long-term effects" and that they "could not determine from the DOD testimony any consistent set of hypothetical Soviet objectives in the strikes analyzed."[9] While the DOD later raised its estimate of fatalities from a Soviet counterforce strike against U.S. ICBM forces to 22 million, it became clear that even revised estimates were far from reliable.[10]

Limited nuclear war could be strategic or tactical. The case that Schlesinger presented was that of a limited counterforce strategic attack. The Office of Technology Assessment has considered a broader range of limited strategic attacks (attacks on the United States that use one hundred-kiloton or larger weapons). Its studies reach similar conclusions. Case 3 in table 3 shows three variations on the Schlesinger case. Such a limited attack would kill from 1 percent to 10 percent of the population, or from $2^1/2$ to 25 million Americans. Case 4 in table 3 takes into account more extensive, yet still limited, strategic attacks. It is limited in the sense that population centers per se are not targeted. At the upper end of limited strategic attacks (excluding targeting of populations), 10 percent to 77 percent fatalities would be expected, or 25 to 200 million deaths.

Many strategists consider the distinction between limited and full-scale nuclear war to be significant. While there may be some militarily significant differences, the effect on populations is much the same. McNamara defined the loss of 20 to 25 percent of the U.S. population as assured destruction. In his studies of limited strategic attacks against the United States, Katz has shown that unless fewer than 100 one-megaton and 200 to 300 one hundred-kiloton weapons are involved, realistic scenarios suggest American fatalities will be higher than McNamara's range for assured destruction.[11]

A limited tactical nuclear war is also possible. While a tactical nuclear war could be fought in any region outside the U.S. and Soviet land masses, the highest concentration of tactical nuclear weapons is in Europe where NATO and Warsaw Pact forces have over 10,000 tactical or

---

7. U.S. Senate, Committee on Foreign Relations, Subcommittee on Arms Control, International Organizations and Security Agreements, *Analyses of Effects of Limited Nuclear Warfare* (Washington, D.C.: Govt. Print. Off., 1975), p. 113.
8. Ibid., p. 115.
9. Ibid., p. 4.
10. Ibid., p. 45.
11. Arthur Katz, *Life after Nuclear War: The Economic and Social Impacts of Nuclear Attacks on the United States* (Cambridge, Mass.: Ballinger, 1982), p. 103.

Table 3.

Fatality Estimates for Various Nuclear Attacks

Case	Office of Technology Assessment Attack Cases	Population Posture	Percentage of National Fatalities Low range	High range
2	{ Small attack on U.S.	(not available)		
	{ Small attack on U.S.S.R.	(not available)		
3	⌠ Attack on U.S. ICBMs	In place	1–3	8–10
	⎮ Attack on Soviet ICBMs	In place	1	1–4
	⎮ Attack on U.S.	In place	1–5	7–11
	⎮ counterforce targets	Evacuation	—	5–7
	⎮ Attack on Soviet	In place	1	1–5
	⌡ counterforce targets	Evacuation	(1–2)*	
4	⌠ Attack on U.S. counterforce	In place	35–50	59–77
	⎮ targets, other military targets,	Evacuation	10–26	32–43
	⎮ and economic targets			
	⎮ Attack on Soviet counterforce	In place	20–32	26–40
	⎮ targets, other military targets,	Evacuation	(9–14)*	
	⌡ and economic targets			
3 (excursion)	⌠ Attack on U.S. counterforce	In place	14–23	26–27
	⎮ targets and other military targets	Evacuation	—	18–25
	⎮ Attack on Soviet counterforce	In place	15–17	22–24
	⌡ targets and other military targets	Evacuation	(6–9)*	
4 (excursion)	⌠ Attack on U.S. counterforce	In place	—	60–88
	⎮ targets, other military targets,	Evacuation	28–40	47–51
	⎮ economic targets and			
	⎮ population			
	⎮ Attack on Soviet counterforce	In place	—	40–50
	⎮ targets, other military targets,	Evacuation	(22–46)*	
	⎮ economic targets and			
	⌡ population			

*Percentages refer to fatality estimate if partial evacuation of the population has occurred.
Source: Adapted from Office of Technology Assessment, The Effects of Nuclear War (Washington, D.C.: Govt. Print. Off., 1979), p.140.

theater nuclear weapons. Many strategists also believe that Europe is the most likely location for the outbreak of tactical nuclear war. On the basis of these assumptions, Arkin, von Hippel, and Levi tried to determine the effect of such a war.[12] First, a limited tactical nuclear war in Europe would likely involve preemptive strikes against nuclear forces themselves or battlefield exchanges of nuclear weapons or both. Second, assuming the conflict stayed in the Germanies and did not include the targeting of civilians, for preemptive strikes there would be at least 172 targets (composed of 92 military airbases, 56 surface-to-surface missile sites, and 24 known or suspected nuclear weapons storage depots). Third, the detonation of a 200-kiloton warhead at two kilometers above

12. William Arkin, Frank von Hippel, and Barbara Levi, "The Consequences of a 'Limited' Nuclear War in East and West Germany," in The Aftermath: The Human and Ecological Consequences of Nuclear War, edited by Jeannie Peterson (New York: Pantheon, 1983), p. 180.

each of these targets would destroy it. Fourth, a battlefield exchange might plausibly involve up to 1000 tactical nuclear weapons with yields ranging from .1 to 100 kilotons.

A limited preemptive tactical nuclear war involving only East and West Germany and no countervalue strikes would, given the previous assumptions about targets and weapons, have the following results. Since each of the 200-kiloton weapons would be lethal within an area equivalent to 180 square kilometers (70 square miles), 10 percent of the land mass of East and West Germany would fall within the contours of lethality. Since calculations for population density suggest each 200 kiloton airburst would kill 40,000 people in areas of average population density and 500,000 in average urban areas, and given the targets in this scenario and corresponding population density, there would be 1 to 40 million fatalities, with 10 million killed and 10 million seriously injured as a likely estimate. To this total, one could add an additional million fatalities for a battlefield exchange.

What these numbers suggest is that while a limited tactical nuclear war, if contained, would be the least destructive plausible form of nuclear war, even it would likely reach the level of "megadeath"—the death of millions of people.

The Subcommittee on Arms Control, International Organizations, and Security Agreements of the Senate Committee on Foreign Relations (1975 and 1976) was responsible for two major works on limited war. The first, *Analyses of the Effects of Limited Nuclear Warfare*, is a collection of reports. The second, *Effects of Limited Nuclear Warfare*, is the transcript of and addenda to the subcommittee's hearings. These documents provide an unusually detailed and candid look at how defense and civilian experts go about calculating the consequences of various nuclear attacks. Congress, in an effort to get an even more thorough and comprehensive study of the effects of small-scale to large-scale nuclear war, had the Office of Technology Assessment prepare *The Effects of Nuclear War*. This report, which reaches essentially the same conclusions as the Ad Hoc Panel (see *Analyses of the Effects of Limited Nuclear Warfare*), is one of the most frequently cited studies on the consequences of nuclear war.

General Long-Term Consequences

Since many efforts to determine the probable effects of limited and full-scale nuclear war focus on the initial days and months after an attack, long-term consequences have become a distinct, yet highly important, topic. During the late 1970s and early 1980s, the major concern was ozone depletion, and currently the prospect for nuclear winter is receiving primary attention (see subsections on these topics in the next section of this chapter). Nuclear winter receives attention even when limited nuclear wars are considered because nuclear winter may be triggered by the detonation of only a few hundred nuclear weapons. The once controversial but now standard source on the diverse long-term consequences is the report of the Committee to Study the Long-Term Worldwide Effects of Multiple Nuclear Weapons Detonations. This research done under the auspices of the National Academy of Sciences considered up to a 10,000-megaton exchange, a full-scale nuclear war (but far from a depletion of nuclear weapon stockpiles).

Economic and Social Consequences

The economic cost and social trauma resulting from nuclear war would be great. The range in estimates depends on many assumptions. Perhaps the most optimistic official estimate forecasts recovery of two-thirds of the preattack gross national product within nine years, followed by a 4 percent yearly growth rate.[13] Some writers predict a return to medieval-style economics.[14] Arthur Katz has contributed perhaps more than anyone else to the credible projection of economic and social consequences. Katz authored the report, *Economic and Social Consequences of Nuclear Attacks on the United States*, for the Senate's Committee on Banking, Housing and Urban Affairs. Katz converted this report into a much larger book, *Life after Nuclear War*. These two sources, especially the latter, are among the most detailed and comprehensive treatments of economic and social issues.

Other Consequences

Studies of physical, especially incendiary, effects of nuclear war include Chandler et al., Lewis, Martin, and Wiersma and Martin. Among noteworthy environmental and ecological studies are Ayres, Stockholm International Peace Research Institute, United Nations (1979), and Woodwell. Studies focused on medical and radiological effects include Ervin, Rotblat, United Nations Scientific Committee on the Effects of Atomic Radiation, and United States (1980). For studies on the social effects see Bergstrom et al., Nordlie, Nordlie and Popper, Smelser, and United Nations (1968).

Summary

Typically, these studies conclude that the human species probably will survive even full-scale nuclear war and that some of the survivors will be in the nations involved in the nuclear war. However, a nuclear

---

13. U.S. Department of Defense, *DCPA Attack Environment Manual*, Chapter 8, Panel 18.

14. See for example J. Carson Mark, "Consequences of Nuclear War," in *The Dangers of Nuclear War*, edited by Franklyn Griffiths and John C. Polanyi (Toronto: Univ. of Toronto Pr., 1979).

war would kill about half the U.S. population and about half the Soviet population, as well as tens of millions of persons in adjacent nations. Recovery would take a decade or more, and other nations would likely emerge as the global powers. These conclusions about the likelihood of some survivors and some recovery have led to many specialized studies, often designed to provide data on how to further enhance survival and recovery.

Possible Catastrophic Consequences of Nuclear War

The blast damage of the global supply of nuclear weapons could destroy only a very small percentage of the earth's surface. Fallout, on the other hand, could blanket much of the planet following a major nuclear exchange. While the prospect of all humans dying from fallout may once have been a widely accepted belief, it has never had serious scientific support. In fact, the more catastrophic forecasts are not based on levels of radiation per se. For example, to blanket the earth with lethal radiation would require the detonation of about one million megatons of fissionable material.[15] Since the global total is many times less than that (perhaps fifty times less), fallout alone would not spell a global catastrophe that would be fatal for all humans.

The more serious issues that have been debated over the last several years include the effects of electromagnetic pulse (EMP), the effects of ozone depletion, the medical consequences, and nuclear winter. The first two, EMP and ozone depletion, are probably now viewed as less serious than the latter two, the medical and environmental consequences.

Electromagnetic Pulse

Awareness of the very negative impact of electromagnetic pulse (EMP) occurred as a result of the postwar atmospheric tests, particularly those that were conducted at very high altitudes. When the third edition of *The Effects of Nuclear Weapons* was published in 1977, it included for the first time a chapter on electromagnetic pulse. Following atmospheric nuclear tests, electrical equipment at considerable distances from the detonations malfunctioned. Eventually, EMP was isolated as the cause of these equipment failures. The concern about the consequences of EMP in nuclear war is not over any damage to persons or the environment (which would be negligible). Rather, EMP threatens the civilian and military technology, especially the electrical and electronic systems

---

15. Adams and Cullen, *The Final Epidemic*, pp. 115–16.

that support communication, command, and control.[16] Computers, of course, are quite susceptible, and the warning, waging, and ending of nuclear war are closely tied to computers. So, indirectly, since it could prevent communication and lead to the loss of control, EMP suggests that even limited nuclear use could escalate and the major catastrophe of full-scale nuclear war could result.

The need to modify susceptible equipment is not restricted geographically since as the height of a nuclear detonation increases, so does the area affected by EMP. For example, whereas the EMP of a surface blast probably would not lead to equipment malfunction beyond ten kilometers from the detonation, the EMP of an airburst at eighty kilometers would affect an area one thousand kilometers in radius.[17]

Broad's series of articles popularize EMP as the "chaos factor" in nuclear planning. Since any potential adversary in a nuclear war might use very high airbursts to disrupt C3I (communication, command, control, intelligence) during an attack, the focus in this debate has been on the threat to C3I and the means to preserve it long enough for nuclear weapons to be launched in retaliation. That was the goal of Ivy League (the SIOP test conducted in 1982 and discussed in Chapter 4). Successful use of EMP by an adversary is often termed decapitation—the opponent loses the C3I systems necessary to wage nuclear war.[18] Contrasting approaches to EMP can be found in Fitts and Ball. Important governmental investigations on EMP include Committee on Armed Forces, Congressional Budget Office, and Department of Defense.

Ozone Depletion

Until concern about medical consequences and nuclear winter captured media attention, the prospect of serious ozone depletion was regarded by many as the most catastrophic consequence of nuclear war. Since the mid-1970s scientists had been warning that the nitrous oxides released after nuclear detonations could rise to the upper atmosphere and deplete a large proportion of the ozone layer that protects most organisms. Even moderate reduction in the ozone layer would have severe repercussions. Organisms, including humans, would be exposed to much higher levels of ultraviolet radiation causing severe sunburns and blinding of animals globally. Moreover, climatic changes would also occur. Many experts predicted a slight, though temporary, drop in temper-

16. Samuel Glasstone and Philip J. Dolan, eds., *The Effects of Nuclear Weapons*, 3rd ed. (Washington, D.C.: Govt. Print. Off., 1977), pp. 514ff.

17. Report of the Secretary General of the United Nations, *Nuclear Weapons* (Brookline, Mass.: Autumn Pr., 1980), p. 200.

18. Popular discussions of this problem include the chapter on "The Headless Dragon" in Nigel Calder, *Nuclear Nightmares: An Investigation into Possible Wars* (New York: Penguin, 1979), and the chapter (and bibliography) "Decapitation" in Peter Pringle and William Arkin, *SIOP: The Secret U.S. Plan for Nuclear War* (New York: Norton, 1983).

ature globally. A global cooling of only a few degrees would dramatically and negatively affect agriculture. Given these prospects, some writers presented human extinction as a possible consequence of ozone depletion and, hence, as a compelling reason to avoid nuclear war.

The magnitude of ozone depletion that is predicted, however, is closely tied to the type of nuclear war that one assumes. For large amounts of nitrous oxides to reach the upper atmosphere, nuclear detonations in excess of one megaton are required. At one time both the United States and the Soviet Union had many multimegaton nuclear weapons in their arsenals. Nevertheless, because of major advances in missile accuracy and stress on counterforce strategies, even strategic nuclear weapons now increasingly have submegaton yields. Moreover, more recent scientific studies have scaled down somewhat the effects that would result even if large amounts of nitrous oxides reached the upper atmosphere.

The initial warning of the potential for ozone depletion came from the National Academy of Sciences' Committee to Study the Long-Term Worldwide Effects of Multiple Nuclear Weapons Detonations. Schell is perhaps the major popularizer of the threat posed by ozone depletion. Nevertheless, even before his book was published the United Nations (1980) reanalyzed the reference scenario of the National Academy of Sciences and concluded that it did not correspond well with current arsenals and strategies.

Discussion of ozone depletion in light of even more recent scientific studies can be found in the hearings on *The Consequences of Nuclear War on the Global Environment* conducted by the Subcommittee on Investigations and Oversights of the Committee on Science and Technology of the U.S. House. Both the updated analysis by the National Academy of Sciences and the reference scenario used by the Ambio Advisory Group are cited and similar conclusions are drawn: restoration of ozone to former levels will likely occur in two rather than four years, and the percentage of temporary depletion could be as low as a few percent.[19] Even in those hearings, while radiation, electromagnetic pulse, and ozone depletion are addressed, the issues of medical consequences and what has subsequently come to be referred to as nuclear winter receive greater attention.

Medical Consequences

From a medical point of view, the effects of nuclear war would be unmanageable if any population centers are hit. For example, if a single

19. U.S. House, Committee on Science and Technology, Subcommittee on Investigations and Oversights. *The Consequences of Nuclear War on the Global Environment* (Washington, D.C.: Govt. Print. Off., 1983), especially remarks by James Friend on p. 113 and John Birks on p. 125.

strategic nuclear weapon hit a major city, there would be more burn victims requiring elaborate care than there are burn beds in the entire United States. Almost all severely burned victims would die, and their number increases by the thousands with each population center that is hit. Since most medical facilities and physicians are located in urban areas, the destruction of cities will reduce even more the only marginal medical aid that could be offered. Moreover, burn victims will account for only one of the major categories of casualties from each nuclear detonation in a populated area. Each population area will have large numbers of people with broken and crushed bones, lacerations, ruptured eardrums and lungs, loss of vision, and radiation sickness. There will be many human and animal corpses, and a multitude of infections and communicable diseases will soon ravage the survivors who, in addition to all of the above conditions, will be under great and long-lasting psychological stress. Despite these facts, the debate on whether and how to respond to such medical crises has been going on for decades, and three basic positions have emerged.

The first position is that these medical crises cannot be managed. Many in the medical profession contend that a major nuclear war would lead to such severe medical crises that social, perhaps even biological, survival could not be attained by the nations involved. They feel that the surviving medical facilities and personnel will make little difference in efforts to recover. Some consider it illusory to plan for postattack recovery and immoral to participate in such planning as it may make nuclear war appear to be survivable and a viable option in a major international crisis. This viewpoint is argued most conspicuously by the medical groups, Physicians for Social Responsibility and International Physicians for the Prevention of Nuclear War.

The second position is that these medical crises can be managed if civil defense programs are adequate. Physicians have been quite vocal in their criticisms when testifying at congressional hearings on the consequences of nuclear war and on civil defense; government officials have been placed on the defensive regarding their plans. The basic crisis management documents on which these officials rely are often rather sparse in providing reliable data to support claims about the nature of the postattack environment and how it can be managed. Nevertheless, these officials argue that civil defense will significantly enhance survival and that efforts to provide medical aid can and should play a role in planning for recovery.

In contradistinction to these two groups stand the survivalists, who argue the third major position. Against the physicians, they agree with the government that nuclear war will be manageable for the survivors, but, against the government, they disagree that management should be under the authority of local, state, and national officials. Like the physicians, survivalists are skeptical of officials plans, but in defiance of governmental plans, survivalists plan to manage on their own. They believe that individuals can stockpile the numerous items needed for survival, including medical supplies and firearms. Some survivalists relocate to less targeted areas in the United States or even other countries. In a postattack world, they believe the individual does not owe government any political obligation and should not feel compelled to aid others with whom they come into contact. Many survivalists, then, would be political anarchists and ethical egoists in a postattack environment.

For arguments by advocates of the first position that nuclear war is medically unmanageable, see especially Adams and Cullen, and also Chivian and Chivian, and Leaning and Keyes. For the presentation of the governmental view that medical and other related crises can be managed, see the governmental sources cited in the first two sections of this chapter and those on civil defense cited in Chapter 3. For the survivalist position, the key text is Clayton.

Nuclear Winter

The most recent, and largely one-sided, debate concerns the prospect for nuclear winter following nuclear war. Of all the potential catastrophic consequences of nuclear war, nuclear winter most strongly points toward the prospect for biocide—nuclear winter could destroy all life on the planet. Nuclear winter is not a function of radiation, EMP, ozone depletion, or medical crises. Rather, it is related to the blast and heat of nuclear detonations. Following nuclear detonations on or near the earth's surface, massive fires are likely. Forests and cities with their numerous combustible fuels will burn and produce enormous amounts of smoke, soot, dust, and other particles. As this material is taken up into the atmosphere, it will form massive dark clouds that eventually will blanket much, if not all, the earth's atmosphere.

For a couple of weeks only a few percent of the average amount of sunlight will reach earth and temperatures may drop to between $-15°$ and $-25°$ centigrade. Subfreezing temperatures would last for months, wiping out many needed agricultural efforts. Wheat and other crops, along with livestock, would be largely destroyed. Loss of fuels, fertilizers, pesticides, feeds, hybrid seeds, and human laborers would greatly hinder agricultural efforts. Ecosystems would suffer major disruptions for at least a year. Numerous species would become extinct, and survival would be precarious in both the Northern and Southern Hemispheres. These results are summarized in figure 5.

A few books address nuclear winter. *Ambio* magazine has released *The Aftermath* (which is a reprint of the initial research on nuclear winter that appeared in *Ambio* in June 1982), and Harwell has published *Nuclear Winter*. See too Greene et al. and Grinspoon. Articles by Turco et

Figure 5.
Timing and Magnitude of Effects of a Large-Scale Nuclear War.

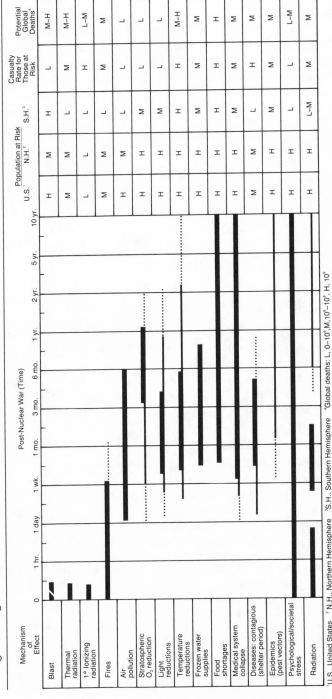

¹U.S., United States   ²N.H., Northern Hemisphere   ³S.H., Southern Hemisphere   ⁴Global deaths: L, 0–10⁶; M, 10⁶–10⁸; H, 10⁹

Source: Mark A. Harwell, *Nuclear Winter: The Human and Environmental Consequences of Nuclear War* (New York: Springer-Verlag, 1984), p. 154.

al. and Ehrlich et al. presented the findings reached by the research teams associated with Harwell and Sagan. Also helpful are the House hearings (September 15, 1982) on *The Consequences of Nuclear War on the Global Environment*. Similar conclusions regarding nuclear winter have been reached in the Soviet Union; see Aleksandrov and Stenchikov. Ehrlich et al. have also published the proceedings of one of their conferences.

Conclusion

There are many very serious effects of nuclear war. Differing life-threatening consequences have to be faced from the moment of detonation, some for at least a decade. Whereas blast, radiation, and fires are the major initial threats, factors like ozone depletion, temperature reduction, and contagious diseases would be the major problems after the first week and for the next several months. Some factors, like food shortages and psychological stress, would persist for many years. Figure 5 includes many of these factors along with nuclear winter.

Bibliography

The following bibliography includes references cited in the text and related suggested readings; references are organized according to the major divisions of the chapter. The full bibliographic reference for all sources cited in the text of this chapter may be found in either the annotated or unannotated bibliography corresponding to the division of the text in which the source is cited. Sources that are annotated are the most essential or significant.

Nuclear Weapons Tests

Berkhouse, L., et al. *Operation Crossroads–1946*. Prepared by the Defense Nuclear Agency. Springfield, Va.: National Technical Information Service, 1984.

One of a series of volumes authorized by the Department of Defense to provide data to the Centers for Disease Control and Veterans Administration on radiation exposure of persons participating in the 235 atmospheric nuclear weapon tests that took place between 1945 and 1962. Crossroads was the first postwar nuclear weapons test series. It was conducted at Bikini Atoll; Able, a twenty-three–kiloton airburst, occurred on July 1, 1946; and Baker, a twenty-three–kiloton underwater detonation, occurred on July 25, 1946. This report provides an incredibly detailed record of the 42,000 personnel, 251 ships, and 156 aircraft involved in what was then the largest U.S. peacetime military operation.

The Committee for the Compilation of Materials on Damage Caused by the Atomic Bombs in Hiroshima and Nagasaki. *Hiroshima and Nagasaki: The*

Physical, Medical, and Social Effects of the Atomic Bombings. Translated by Eisei Ishikawa and David L. Swain. New York: Basic Books, 1981. The most authoritative study of the effects of the atomic bombings of Hiroshima and Nagasaki. Three editors, along with thirty-four specialists in physics, medicine, social sciences, and humanities, analyzed the cumulative literature on the topic. The bibliography for this literature runs nearly forty pages in the book. This book supersedes that literature and is indispensable for a thorough understanding of the consequences of using nuclear weapons. In addition to providing a wealth of information in a quite readable style, the book often makes its political point about the unacceptability of nuclear weapons.

Gladeck, F. R., et al. *Operation Ivy: 1952.* Prepared for the Defense Nuclear Agency. Springfield, Va.: National Technical Information Service, 1982.

   Like *Operation Crossroads* (above), *Operation Ivy* provides a wealth of information on an important nuclear weapons test series. The first of these two detonations at Enewetak Atoll was the first hydrogen bomb test. Mike, the code name, was detonated on November 1, 1952, and was a 10.4-megaton blast. The blast destroyed the island on which the bomb was detonated and left a submerged crater 6,300 feet in diameter and 160 feet deep.

Glasstone, Samuel, et al., eds. *The Effects of Atomic Weapons.* Washington, D.C.: Govt. Print. Off., 1950.

   Published in 1948 and revised in 1950, this book, prepared largely by the Los Alamos Scientific Laboratory, was the first major study of the effects of atomic or fission weapons. It predates the development of the hydrogen or fusion bomb and, hence, uses the narrower term "atomic" rather than the broader term "nuclear." However, it is based on data from not only Trinity, Hiroshima, and Nagasaki, but also early postwar tests. Contains one or more chapters on the following topics: atomic explosions; shock; physical, thermal, radiological, and incendiary effects; decontamination; and protection. Though of historical interest, this book was superseded by *The Effects of Nuclear Weapons* (below).

_____ and Philip J. Dolan, eds. *The Effects of Nuclear Weapons.* Rev. U.S. Department of Defense and U.S. Department of Energy. Washington, D.C.: Govt. Print. Off., 1977.

   The standard, though technical, source on the subject. Originally published in 1957 in order to cover the effects of hydrogen or fusion in addition to atomic or fission weapons, this book superseded *The Effects of Atomic Weapons* (above). The book has undergone periodic revision as new data has been learned, especially from high altitude tests that led to the addition of a chapter on electromagnetic pulse. Contains one or more chapters on nuclear explosions, blast, shock, thermal and nuclear radiation, electromagnetic pulse, and biological effects. Each chapter has a bibliography, and the book has both a glossary and a "nuclear bomb effects computer."

Japan National Preparatory Committee. *A Call from Hibakusha of Hiroshima and Nagasaki: Proceedings of International Symposium on the Damage and After Effects of the Atomic Bombings of Hiroshima and Nagasaki.* Tokyo: Asahi Evening News, 1978.

Before the publication of *Hiroshima and Nagasaki* (above), an important source for documents on physical, medical, and social effects of atomic bombings of Hiroshima and Nagasaki. Much attention is devoted to preventing further use of nuclear weapons and to eventual nuclear disarmament. The symposium, of which this book is the record, was held in Tokyo, Hiroshima, and Nagasaki between July 21 and August 9, 1977, and brought together many international experts and peace activists.

Lifton, Robert Jay. *Death in Life: Survivors of Hiroshima*. New York: Random, 1967.

The first major study of the psychological effects of the atomic bombing of Hiroshima. Relies on the few studies that have been done but draws mostly from interviews with over seventy survivors, some randomly selected and others specifically selected because of their articulateness and work with atomic bomb problems. Many of the chapters consist largely of quotations from survivors. Lifton provides a composite view, indicating specific problems with which survivors had to deal and various ways survivors did respond. As a psychiatrist, Lifton views the response as neither pathological nor normal but adaptive.

Bauer, E. and F. R. Gilmore. "Effect of Atmospheric Nuclear Explosions on Total Ozone." *Reviews of Geophysical and Space Physics* 13 (1975): 451–58.

Chivian, Eric and Joseph Goodman. "What Soviet Children Are Saying about Nuclear War." *Report* (Newsletter of the International Physicians for the Prevention of Nuclear War) 2, no. 1 (Winter, 1984): 10–12.

Conrad, Robert A., et al. *Twenty-Year Review of Medical Findings in a Marshallese Population Accidentally Exposed to Radio-active Fallout*. Publication No. BNL50424. Upton, N.Y.: Brookhaven National Laboratory, 1975.

Hines, Neal O. *Proving Ground: An Account of the Radiobiological Studies in the Pacific, 1946–1961*. Seattle: Univ. of Washington Pr., 1962.

Johnston, Harold, Gary Whitten, and John Birks. "Effect of Nuclear Explosions on Stratospheric Nitric Oxide and Ozone." *Journal of Geophysical Research* 78 (1973): 6,107–35.

Nagai, Takashi. *We of Nagasaki: The Story of Survivors in an Atomic Wasteland*. Translated by I. Shirato and H. B. L. Silverman. New York: Duell, 1951.

Schultz, V. "References on Nevada Test Site Ecological Research." *Great Basin Naturalist* 26 (1966): 79–86.

U.S. Strategic Bombing Survey. 10 vols. New York: Garland, 1976.

Wasserman, Harvey and Norman Solomon. *Killing Our Own: The Disaster of America's Experience with Atomic Radiation*. New York: Delacorte, 1982.

Probable Consequences of Nuclear War

Committee to Study the Long-Term Worldwide Effects of Multiple Nuclear-Weapons Detonations, Assembly of Mathematical and Physical Sciences, National Research Council. *Long-Term Worldwide Effects of Multiple Nuclear-Weapons Detonations*. Washington, D.C.: National Academy of Sciences, 1975.

First major report to concentrate on danger of ozone depletion from nitrous oxides released after nuclear detonations. Considers up to 10,000-megaton exchange yet still affirms survival of biosphere and species even in such a nuclear war. Treats long-term worldwide consequences on atmosphere and climate, natural terrestrial ecosystems, agriculture and animal husbandry, aquatic environment, and somatic and genetic effects on humans.

Katz, Arthur M. *Life after Nuclear War: The Economic and Social Impacts of Nuclear Attacks on the United States.* Cambridge, Mass.: Ballinger, 1982.

About a third of this book is drawn from the author's study, *Economic and Social Consequences of Nuclear Attacks on the United States,* published by Senate Committee on Banking, Housing, and Urban Affairs (below). This book is much more detailed and comprehensive than the Senate study and equally pessimistic. While the first part goes over common data on physical effects, the other two parts focus on Katz's own major contributions to the understanding of the economic and social effects of limited and full-scale nuclear war. He examines such vital areas as food production, energy, supplies, medical care, higher education, psychological crises, social disorganization, and disrupted political authority. This quite readable yet lengthy book is aided by well over a hundred charts, tables, and graphs.

U.S. Congress. Office of Technology Assessment. *The Effects of Nuclear War.* Washington, D.C.: Govt. Print. Off., 1979.

This frequently cited study was prepared as a more thorough and comprehensive study than the earlier *Analyses of Effects of Limited Nuclear Warfare* (below). Reaches quite negative conclusions—even limited nuclear war would inflict unprecedented loss of life and damage to the economic and political system. Has chapters on the effects of single bombs, civil defense, various attack scenarios, and long-term effects. Appendixes include a helpful glossary and a bibliography with sections on physical effects, economic impact, social and psychological factors, and civil defense.

————. Senate. Committee on Banking, Housing, and Urban Affairs. *Economic and Social Consequences of Nuclear Attacks on the United States,* Washington, D.C.: Govt. Print. Off., 1979.

Arthur Katz prepared this study and later incorporated most of it into *Life after Nuclear War* (above). The conclusions are largely negative: significant loss of life and economic assets would follow even a modest nuclear attack. In addition to information on the general effects of nuclear war, the study takes a detailed look at the effects on food production, energy supply, medical care, and higher education. In the context of civil defense (often evacuation), the study investigates likely psychological crises, social disorganization, and disrupted political authority in surviving postattack communities. Over fifty charts, tables, and graphs are provided to illustrate and support the study's findings.

————. Committee on Foreign Relations. Subcommittee on Arms Control, International Organizations, and Security Agreements. *Analyses of the Effects of Limited Nuclear Warfare.* Office of Technology Assessment report prepared for the Subcommittee. Washington, D.C.: Govt. Print. Off., 1975.

In response to questions about then-Secretary of Defense Schlesinger's low estimate of fatalities from limited Soviet counterforce strike (the September 11, 1974, testimony in which he made these claims is addended), the subcommittee collected reports and letters from an ad hoc panel for the Office of Technology Assessment on nuclear effects and from the Department of Defense. The DOD raises its casualty figures, but the ad hoc panel still judges the DOD calculations to be deficient. Much technical, yet illuminating, data.

_____. *Effects of Limited Nuclear Warfare.* Washington, D.C.: Govt. Print. Off., 1976.

The actual hearing that followed collection of reports published as *Analyses of Effects of Limited Nuclear Warfare* (above). Includes testimony from Sidney D. Drell, Richard L. Garwin, and James V. Neel.

Ayres, Robert U. *Environmental Effects of Nuclear Weapons.* Parts I, II, III. HI-518-RR. Harmon-on-Hudson, N.Y.: Hudson Research Institute, 1965.

Bergstrom, S., et al. *Effects of a Nuclear War on Health and Health Services.* Report of the International Committee of Experts in Medical Sciences and Public Health. WHO Pub. A36.12. Geneva: World Health Organization, 1983.

Bensen, David W. and Arnold H. Sparrow, eds. *Survival of Food Crops and Livestock in the Event of Nuclear War.* Washington, D.C.: U.S. Atomic Energy Commission, 1971.

Billheimer, John W. and Arthur W. Simpson. *Effects of Attack on Food Production to Relocated Populations.* Vols. I and II. Washington, D.C.: SYSTAN, 1978. Prepared for the U.S. Defense Civil Preparedness Agency.

Bolt, Bruce A. *Nuclear Explosions and Earthquakes.* San Francisco: Freeman, 1976.

Brown, S. L., H. Lee Mackin, and K. D. Moll. *Agricultural Vulnerability to Nuclear War.* Washington, D.C.: Stanford Research Institute, 1973. Prepared for the U.S. Defense Civil Preparedness Agency.

Chandler, C. C., T. G. Storey and C. D. Tangren. *Prediction of Fire Spread Following Nuclear Explosions.* Research Paper No. PSW-5. Washington, D.C.: U.S. Forest Service, 1963.

Ervin, F. R., et al. "Medical Consequences of Thermonuclear War I. Human and Ecological Effects in Massachusetts of an Assumed Thermonuclear Attack on the United States." *New England Journal of Medicine* 266 (1962): 127-37.

Greene, Owen, et al. *London after the Bomb: What a Nuclear Attack Really Means.* New York: Oxford Univ. Pr., 1982.

Haeland, C. M., C. V. Chester, and E. P. Wigner. *Survival of the Relocated Population of the U.S. after a Nuclear Attack.* Washington, D.C.: U.S. Defense Civil Preparedness Agency, 1976.

Lewis, K. "The Prompt and Delayed Effects of Nuclear War." *Scientific America* 241 (1979): 35-47.

Nordlie, Peter. *Approach to the Study of Social and Psychological Effects of Nuclear Attack.* Report No. HSR-RR-6313-Rr. McLean, Va.: Human Sciences Research, 1963.

_____ and Robert D. Popper. *Social Phenomena in a Post-Nuclear Attack Situation: Synopses of Likely Social Effects of the Physical Damage.* Arlington, Va.: Human Sciences Research, 1961.

Rotblat, Joseph. *Nuclear Radiation in Warfare.* Cambridge, Mass.: Oelgeschlager, Gunn, & Hain, 1981.

Smelser, Neil J. *Theories of Social Change and the Analysis of Nuclear Attack and Recovery.* Report No. HSR-RR-6711-ME. McLean, Va.: Human Services Research, 1967.

Stockholm International Peace Research Institute. *Weapons of Mass Destruction and the Environment.* London: Taylor & Francis, 1977.

Stonier, T. *Nuclear Disaster.* New York: Meridian, 1964.

United Nations. *Effects of the Possible Use of Nuclear Weapons and the Security and Economic Implications for States of the Acquisition and Further Development of These Weapons.* New York: United Nations, 1968.

_____. *The Effects of Weapons on Ecosystems.* New York: United Nations Environmental Programme, 1979.

_____. *Nuclear Weapons.* Brookline, N.Y.: Autumn Pr., 1980.

United Nations Scientific Committee on the Effects of Atomic Radiation. *Ionizing Radiation: Levels and Effects.* 2 vols. New York: United Nations, 1972.

U.S. Congress. Senate. Committee on Labor and Human Resources. Subcommittee on Health and Scientific Research. *Short- and Long-Term Health Effects on the Surviving Population of a Nuclear War.* Washington, D.C.: Govt. Print. Off., 1980.

U.S. Defense Civil Preparedness Agency. Department of Defense. *DCPA Attack Environment Manual.* Washington, D.C.: Govt. Print. Off., 1973.

Wiersma, S. J. and S. B. Martin. *Evaluation of the Nuclear Fire Threat to Urban Areas.* Washington, D.C.: Defense Civil Preparedness Agency, 1973.

Woodwell, G. W., ed. *The Ecological Effect of Nuclear War.* Springfield, Va.: Brookhaven National Laboratory, 1963.

Possible Catastrophic Consequences of Nuclear War

Adams, Ruth and Susan Cullen, eds. *The Final Epidemic: Physicians and Scientists on Nuclear War.* Chicago: Educational Foundation for Nuclear Science, 1981.

A widely cited and rather comprehensive set of articles on the costs and causes of the nuclear arms race and on the consequences and prevention of nuclear war. The book is closely associated with the *Bulletin of the Atomic Scientists,* Physicians for Social Responsibility, and to a lesser extent, with International Physicians for the Prevention of Nuclear War.

Ambio Magazine. *The Aftermath: The Human and Ecological Consequences of Nuclear War.* New York: Random, 1983.

Ambio vol. 11, no. 2–3 (1982) reprinted as book edited by Jeannie Peterson. This double issue of *Ambio* focused on human and ecological consequences of a major nuclear war. Includes articles by specialists on arsenals and their effects and on epidemiological, radiological, atmospheric, ecological, agricultural, economic, and behavioral consequences. Two articles have received particular attention: Paul J. Crutzen and John W. Birks's

"The Atmosphere after a Nuclear War: Twilight at Noon" broke ground on the nuclear winter thesis. William Arkin, Frank von Hippel, and Barbara G. Levi's "The Consequences of a 'Limited' Nuclear War in East and West Germany" is also frequently cited.

Clayton, Bruce D. *Life after Doomsday: A Survivalist Guide to Nuclear War and Other Major Disasters.* Boulder, Co.: Paladin, 1980.

This book is based on the assumption that any disaster, including nuclear war, will have survivors and that the more advanced planning an individual does the better chances for survival are—do not count on government, but make your own provisions. Clayton gives data on relative safety of various parts of the country, what kinds of shelter are available, how to stock food and medicine, how to use weapons (firearms), and several appendices.

Ehrlich, Paul R., et al. "Long-Term Biological Consequences of Nuclear War." *Science* 222 (December 23, 1983): 1293–1300.

One of the early statements in a scientific journal of the danger of nuclear winter. Beyond predicting lowered temperature after nuclear war, it focuses on conditions of starvation and near-lethal radiation. Stresses high number of species that would become extinct, possibly even humans.

Harwell, Mark A. *Nuclear Winter: The Human and Environmental Consequences of Nuclear War.* New York: Springer-Verlag, 1984.

This book developed out of Harwell's effort to provide technical support for one of the early studies of nuclear winter (Ehrlich et al., above). After a chapter on how the analyzed nuclear war scenario was selected and on what immediate consequences would follow, Harwell investigates long-term consequences and recovery potential in the first book on the topic of nuclear winter. Attention is given to the major problems of reduced temperatures, reduced light, and descriptions of agriculture and society. Reaches very pessimistic conclusions about recovery and stresses the possibility of human extinction. Provides a helpful and readable final chapter that summarizes the consequences. Also includes a good bibliography on prior studies relevant to research on nuclear winter.

Turco, R. P., et al. "Nuclear Winter: Global Consequences of Multiple Nuclear Explosions." *Science* 222 (December 23, 1983): 1283–92.

Often referred to as "TTAPS" (for authors' initials), one of the early statements in a scientific journal of the danger of nuclear winter. Based on applying data on effects of volcanic eruptions to consequences of nuclear detonations. Focuses on prospect that dust and smoke from nuclear war could lower temperatures to $-15°$ to $-25°C$ for weeks.

Aleksandrov, V. V. and G. L. Stenchikov. "On the Modelling of the Climatic Consequences of Nuclear War." Moscow: USSR Academy of Sciences Computing Centre, 1983.

Ball, Desmond. "Can Nuclear War Be Controlled?" *Adelphi Papers*, No. 169. London: International Institute for Strategic Studies, 1981.

Bridges, J. E. and J. Weyer. "EMP Threat and Countermeasures for Civil Defense Systems." Chicago: Illinois Institute of Technology Research Institute, 1968.

Broad, William J. "Nuclear Pulse (I): Awakening to the Chaos Factor." *Science* (May 29, 1981): 1009–12.
_____. "Nuclear Pulse (II): Ensuring Delivery of the Doomsday Signal." *Science* (June 5, 1981): 1116–20.
_____. "Nuclear Pulse (III): Playing a Wild Card." *Science* (June 12, 1981): 1248–51.
Chivian, Eric and Suzanne Chivian, eds. *Last Aid: The Medical Dimensions of Nuclear War*. San Francisco: Freeman, 1982.
Convey, C., S. H. Schneider, and S. L. Thompson. "Global Atmospheric Effects of Massive Smoke Injections from a Nuclear War: Results from General Circulation Model Simulations." *Nature* 308 (1984): 21–25.
Ehrlich, Paul, et al. *The Cold and the Dark. Proceedings of the Conference on the Long-Term Biological Consequences of Nuclear War*. New York: Norton, 1984.
Fetter, S. and K. Tsipis. "Catastrophic Releases of Radioactivity." *Scientific American* 244 (1981): 41–47.
Fitts, Richard E. *The Strategy of Electromagnetic Conflict*. Los Altos, Calif.: Peninsula, 1980.
Greene, Owen, Ian Percival, and Irene Ridge. *Nuclear Winter: The Evidence and the Risks*. Cambridge: Polity Pr., 1985.
Grinspoon, Lester, ed. *The Long Darkness: Psychological and Moral Perspectives on Nuclear Winter*. New Haven, Conn.: Yale Univ. Pr., 1986.
Holden, Constance. "Military Grapples with the Chaos Factor." *Science* (September 11, 1981): 1228–30.
Leaning, Jennifer and Langley Keyes, eds. *The Counterfeit Ark: Crisis Relocation and Nuclear War*. Cambridge, Mass.: Ballinger, 1984.
MacCracken, M. C. *Nuclear War: Preliminary Estimates of the Climatic Effects of a Nuclear Exchange*. UCRL-89770. Livermore, Calif.: Lawrence Livermore National Laboratory, 1983.
Ramberg, Bennet. *Destruction of Nuclear Energy Facilities in War. The Problem and the Implications*. Lexington, Mass.: Lexington Books, 1980.
Sagan, Carl. "Nuclear War and Climatic Catastrophe: Some Policy Implications." *Foreign Affairs* 62 (1983): 257–92.
Schell, Jonathan. *The Fate of the Earth*. New York: Avon, 1982.
Scheer, Robert. *With Enough Shovels: Reagan, Bush and Nuclear War*. New York: Random, 1981.
Turco, R. P., et al. "The Climatic Effects of Nuclear War." *Scientific America* 251 (1984): 33–43.
U.S. Congress. House. Committee on Armed Services. *Review of Department of Defense Worldwide Communications, Phase I*. Washington, D.C.: Govt. Print. Off., 1971.
_____. Committee on Science and Technology. Subcommittee on Investigations and Oversight. *The Consequences of Nuclear War on the Global Environment*. Washington, D.C.: Govt. Print. Off., 1983.
_____. Senate. Committee on Banking, Housing, and Urban Affairs. *Civil Defense*. Washington, D.C.: Govt. Print. Off., 1979.

U.S. Congressional Budget Office. *Strategic Command, Control, and Communications: Alternative Approaches for Modernization.* Washington, D.C.: Govt. Print. Off., 1981.

U.S. Department of Defense. *C3I Program Management Structure and Major Programs.* Washington, D.C.: Govt. Print. Off., 1980.

CHAPTER 6

Military and Socioeconomic Consequences of the Arms Race

During the past forty years, the superpower rivalry between the United States and the Soviet Union has been characterized by: (1) vacillating political hostility and cooperation; (2) nuclear policies that threaten each other's population and promise destruction of the industrial and military base of the society; (3) an arms race involving the development and stockpiling of conventional and nuclear weapons and the deployment of these weapons, through trade and treaty agreement, throughout much of the world; (4) participation in numerous civil wars and interstate conflicts (often termed proxy wars); and (5) no direct military confrontation. The lack of direct military confrontation, when combined with the massive destruction expected should even a limited nuclear war occur (summarized in Chapter 5), have led many to argue that our nuclear policies work—"to work for peace, prepare for war." Debates over this interpretation are discussed in Chapters 7 and 8.

This chapter examines some of the costs and consequences of the superpower rivalry *even in the absence of nuclear war*. The dominant general perspective on the arms race has been one of political and economic benefit: the arms race has maintained superpower stability while avoiding direct war, and has led to direct and indirect economic growth. Recently, however, this dominant perspective has come under increasing criticism. These criticisms focus on consequences and typically are categorized under the headings of global militarization and socioeconomic development. Global militarization refers to the increased significance of military institutions and expenditures on civil society. In Chapter 1, we summarized some of the long-term trends in military organization and war-fighting. The modern era was characterized by large military organizations, increased costs in maintaining a military posture, and war between superpowers that, while less frequent, was more lethal, involving large numbers of civilians and military casualties. In the nuclear era, we noted, there has been a nuclearization of forces among the superpowers and a heightened emphasis on avoiding war. In addition, since World War II there has been an intensification of long-term trends in militarization, and the emergence of new patterns (for example, permanent war economies) that call into question the traditional views of the costs and benefits of maintaining security.

In the first part of Chapter 6, we examine a number of these military issues and debates: (1) Is the arms race necessary or overkill? (2) What is the relationship between the arms race and arms control? (3) What is the impact of superpower rivalry on global patterns of military expenditures? (4) How has the superpower rivalry affected the international pattern of arms trade and transfer? (5) What is the impact of the arms race on conventional weaponry? and (6) What impact has the arms race had on horizontal proliferation?

The second part of Chapter 6 examines the relation between the arms race and socioeconomic development. The general question addressed here is the degree to which the arms race lessens or threatens national and global development: (1) What effect does the arms race have on the pattern of development in industrialized countries? and (2) What are the socioeconomic, military, and political effects of the arms race on the pattern of development in developing countries?

Global Militarization

Since World War II, a hierarchical structure of military establishments has evolved, a structure based on dominance and interdependence. The split between nuclear and nonnuclear nations, reflecting but not identical to superpower status, has had major consequences on the size, scope, and function of military organizations. A relatively small number of countries dominate the world's nuclear arsenals and attempt to restrict others from developing a nuclear weapon capacity. The dominant members of this nuclear club compete over their arsenals. Research and development activities among a small number of countries shape the quantity and quality of military arsenals throughout the world: in acting to maintain their nuclear monopoly, in competing for political power and in directing the structure of conventional armaments, these superpowers control an interacting nuclear and conventional arms race. The nonnuclear nations are increasingly dependent on nuclear nations for weapons and military protection.

Is the Arms Race Necessary or Overkill?

Chapter 2 traced the quantitative and qualitative evolution of nuclear weapons since World War II. Unlike previous eras when the stockpiling of conventional armaments and the size of uniformed military personnel were seen as the primary indicators of relative strength, nuclear force strength is more determined by the number of viable targets (whether counterforce or countervalue) and the opponent's capacity to retaliate. Chapters 2, 3, and 4 detailed the uneven relationship between the expanding nuclear arsenals of the superpowers and targets. Thus, an emphasis on countervalue targeting, dominant in the early years of the nuclear era, could estimate that a 200-EMT (equivalent megatonnage; see Glossary) attack would destroy up to three-fourths of Soviet industry and one-fourth of the population even when the United States had an arsenal with an EMT capacity of 6,000 megatons (see Enthoven and Smith). Counterforce targeting, while apparently more humane in targeting military capacity, facilitated a quantitative and qualitative expansion of nuclear weapons because of the increased number of targets. Yet the question remains: is further weapon production necessary or redundant? Is it overkill?

Figure 6 illustrates the proliferation of the nuclear arsenals of the United States and the Soviet Union against a backdrop of important dates of weapon and delivery system development. The Soviet Union has rather consistently added to its arsenal, rarely replacing or depleting earlier weapons systems. In contrast, the quantitative buildup of the United States, a fifteenfold increase, occurred between 1955 and 1965. Since this period, the number of weapons has been relatively stable, with new weapons replacing older ones. These more accurate and efficient weapons are smaller in yield (approximately one-half that of the nuclear arsenal of 1960).

Of the approximately 46,000 nuclear weapons possessed by the United States and Soviet Union today, an estimated 17,400 (37.5 percent) are categorized as strategic (capable of reaching the other's homeland). Of this total, the United States is estimated to have 10,000 strategic weapons, with an aggregate explosive power of at least 4,000 equivalent megatons; the Soviet Union is estimated to have 7,400 strategic weapons with an estimated aggregate explosive power of between 6,000 and 7,000 megatons.[1] Fewer than 300 strategic weapons are found in the other nuclear nations: the United Kingdom has 192, France has 80, and China is estimated to have 4.[2] Strategic weapons are seen as the most important elements of a nuclear arsenal and traditionally the focus of arms control efforts. The largest group of nuclear weapons are classified as tactical (to be used in a battlefield or theater situation). Of the estimated 24,000 tactical weapons, approximately 70 percent are in the possession of the United States, with another 26 percent in the arsenal of the Soviet Union. Finally, the Soviet Union is thought to have 3,500

1. Paul P. Craig and John A. Jungerman, *Nuclear Arms Race: Technology and Society* (New York: McGraw-Hill, 1986), p. 6.
2. Ruth Sivard, *World Military and Social Expenditures 1983* (Washington, D.C.: World Priorities, 1983), p. 6.

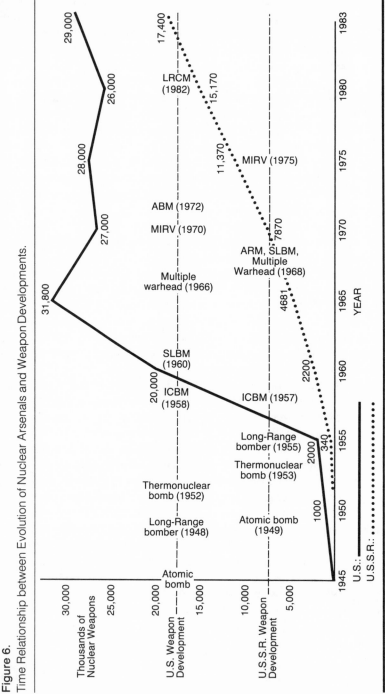

Figure 6.
Time Relationship between Evolution of Nuclear Arsenals and Weapon Developments.

intermediate-range weapons (those having a range of 1,500 miles or more), compared to 1,300 for the United States.[3]

In recent years, both the United States and the Soviet Union have initiated programs to modernize existing systems and expand their nuclear arsenals. In quantitative terms, the United States plans to produce an additional 23,000 nuclear weapons over the next decade. Taking into account weapons to be withdrawn or replaced, it is currently estimated that the United States will equal its 1967 figure of 32,000 nuclear weapons by 1990, increasing the aggregate explosive power of its arsenal by 75 percent of its 1980 level.[4] In terms of quality of its weapons, the United States is in the process of producing and deploying the B-1 bomber, the advanced technology Stealth bomber, and 3,800 cruise missiles on bombers. Sea-launched cruise and ballistic missiles are being deployed on the submarine component of the American strategic TRIAD. MX missiles are being added to the ICBM component of the TRIAD, although the number of missiles and warheads remains in doubt. More sophisticated guidance systems are to be adapted to existing weapon systems, significantly increasing overall efficiency. Finally, as discussed in Chapter 2, the United States plans to conduct a variety of research and testing programs in ballistic missile defense and antisatellite weapons. The Soviet Union is known to be working on development of a solid fuel system (a technology the United States developed long ago) and long-range cruise missiles, and is possibly developing a new bomber.[5] In the absence of negotiated agreements to limit strategic weapons, it is estimated that by 1990 the Soviet Union could increase the number of its delivery vehicles by 25 percent and its aggregate explosive power by 100 percent over 1980 levels.[6]

Table 4 reproduces a recent Department of Defense assessment of the relative strengths and weaknesses of Soviet and American nuclear forces. These estimates are quite conservative and do not take into account developments since 1982. Its inclusion is intended to illustrate the various components of strategic nuclear forces that suggest parity or superiority and the shifting base of competition implicit in the arms race— as one side perceives weakness in its position, pressures emerge to overcome the other's lead. Nothing in the table denotes sufficiency of forces; rather the emphasis is on relative position. Persons interested in comparisons of American-Soviet arsenals should consult Collins; Cochran, Arkin, and Hoenig; International Institute for Strategic Studies; and Sivard.

3. Ibid.
4. Craig and Jungerman, *Nuclear Arms Race*, p. 6.
5. Stockholm International Peace Research Institute (SIPRI), *World Armament and Disarmament: SIPRI Yearbook 1983* (New York: Taylor & Francis, 1983), p. lv.
6. Craig and Jungerman, *Nuclear Arms Race*, p. 6.

Table 4.

Department of Defense Assessment of Relative U.S. and Soviet Technological Levels in Deployed Military Systems

Deployed System	1980 U.S. Superior	1980 U.S. and USSR Equal	1980 USSR Superior	1982 U.S. Superior	1982 U.S. and USSR Equal	1982 USSR Superior
Strategic						
ICBM		x			x	
SSBN/SLBM	x→ *					
SSBN					x	
SLBM				x→		
Bomber	x→			x		
SAM			x			x
BMD			x			x
Antisatellite			x			x
Cruise missile			x			x
Tactical land forces						
SAM (including naval)		x			x	
Tank			←x		x	
Artillery	x→				x	
Infantry combat vehicle			x			x
Antitank guided missile		x			x	
Attack helicopter	x→				x	
Chemical warfare			x			x
Theater ballistic missile		x			x	
Air forces						
Fighter/attack aircraft	x			x→		
Air-to-air missile	x			x		
PGM	x			x→		
Airlift	x			x		
Naval forces						
Nuclear-powered submarine		x			x	
Antisubmarine warfare	x→			x		
Sea-based air	x→			x		
Surface combatant		x			x	
Cruise missile		x		x→		
Mine warfare			x			x
Amphibious assault	x→			x		
C3I						
Communications	x→				x	
Command and control		x			x	
Electronic counter-measures		x			x	
Surveillance and reconnaissance	x→			x→		
Early warning	x→			x→		

*Arrow represents significant shift of advantage.
Source: *SIPRI Yearbook 1984*, p. 92.

Chapters 2 and 3 indicated that the United States and the Soviet Union achieved strategic parity (rough equivalence in delivery vehicles, warheads and aggregate explosive power in the early 1970s. Each had achieved a second-strike capacity (assured destruction) and had aggregate explosive power many times that seen as necessary to maintain minimum deterrence. Yet each side has continued to expand and modernize. Is this overkill? Is such a pattern inevitable?

Carnesale et al. suggests five approaches to this question, which we reduce to three perspectives. The first perspective emphasizes that the arms race continues because of changing security requirements either in extending policy options or in reducing the potential advantage that an opponent may gain by arms production. Examples of extending policy options include the adoption of the policy of massive retaliation in the 1950s, extended deterrence (forces to deter an attack on one's allies rather than one's homeland), and the more recent policy emphasis on "prevailing" in a nuclear war should it occur. The second variant of this perspective emphasizes arms racing on the basis of denial of advantage; the recent emphasis on development of the MX missile to use in bargaining is one example.

A second perspective focuses on the action-reaction cycle seemingly characteristic of the arms race. Two quite different variants may be identified in this perspective. The first stresses the role of uncertainty and misperception: mistaken intelligence reports or unfounded estimates of an opponent's strength or intentions lead to defensive reaction to maintain stability or superiority. The debate over the missile gap in 1960 in which mistaken and unfounded reports of Soviet strength leading to an American weapon buildup and the debate in the 1980s over the window of vulnerability (the misperception of an opponent's intentions) illustrate this view. The second argument holds that the arms race, particularly the action-reaction cycle, is a subset of the more general competition and rivalry between the superpowers: if it can be done, we should do it first. The decision to build the hydrogen bomb illustrates such a view. The third perspective emphasizes technological and internal political pressures on the arms race: security requirements or international political objectives are secondary to either technological innovation or the workings of the military-industrial complex. The emphasis on technology includes the arguments that bureaucratic infighting among the armed services leads to unending adoption of the latest innovations, and that unexpected breakthroughs in nonmilitary research become applicable to weapon systems in a serendipitous effect. Writings on the development of the MIRV capacity illustrate this view. A related view stresses the political and economic power of the defense industry in influencing weapon decisions, as when it argues that a weapon development program is good for the economy and employment.

These perspectives are not exclusive of the others, but focus on the differing dimensions—political, military, economic, and trust—that characterize the arms race. Each has some validity in explaining some aspect of the arms race to this point. But of these perspectives, only changing security requirements appears to assign much legitimacy to further stockpiling of weapons. The question of overkill is a question of policy objectives: how many weapons are necessary to achieve deterrence or superiority? Does weapon development beyond minimum deterrence gain additional security? Even this approach suffers from the dual problem of deciding what constitutes a valid policy option and whether such an option could be met in ways other than more weapon development. Given the costs associated with continued arms racing, it is legitimate to ask why not maintain existing arsenals or reduce them? An important characteristic common to the three perspectives is the degree to which weapons and policies concerning their use often follow separate paths: weapons are developed independent of legitimate security interests, and policies leading to weapon development often have more political than military utility.

To explore the various explanations of the arms race, see Bottoms and Barnet (1982) for critical assessments; Enthoven and Smith; Carlton and Schaerf; and Gray are suggested for examinations of specific weapon decisions; Gillespie et al. (1978, 1979), Hamblin et al., Hill, Moll and Moll, and Ostrom should be consulted for examples of empirical tests of arms racing.

The Arms Race and Arms Control

During the past twenty years, arms control has become an institutionalized part of the arms race between the United States and the Soviet Union. It emerged as a practical response to negative public opinion toward the arms race and as an element in the theory of deterrence. And while public opinion has at various times supported the reduction in the size of the nuclear arsenals, arms control in its relation to deterrence has become a facilitant to further weapon production. Disarmament normally refers to the reduction or elimination of weapons and is predicated on the idea that such weapons are harmful. Within the theory of deterrence, arms control attempts to "reduce or eliminate any incentive that might exist to attack. If the military forces of two potential enemies are such that an attack will yield no advantage to the attacker, peace will be assured."[7] Arms control efforts may be directed toward such matters as improving communications, reducing the possibility of a surprise first strike and more generally, negotiating the size and characteristics of nu-

7. Shelia Tobias et al., *The People's Guide to National Defense* (New York: Morrow, 1982), p. 86–87.

clear arsenals (such as bans or limitations on research, testing, development, deployment, and use of weapons).

The arms race as we have described it has evolved from a primarily quantitative to qualitative orientation. Both dimensions (more weapons or more accurate weapons) may represent threats to the theorized stability of deterrence. Early arms control efforts were directed toward slowing the sheer number of weapons, while more recent efforts have been directed toward reducing the first-strike character of certain weapon systems.

Table 5 presents descriptions of bilateral (between the United States and Soviet Union) and multilateral agreements concerning nuclear weapons. The first group of agreements has generally been successful in reducing horizontal proliferation (although significant disagreements exist over sections of the Outer Space Treaty and the Nonproliferation Treaty). The second category includes agreements to improve communications between the United States and Soviet Union and to reduce the possibility of accidental war. Such agreements appear to have wide support but are difficult to evaluate in terms of effectiveness. The third category includes the multilateral Partial Test Ban Treaty that was negotiated in response to heightened public awareness and international concern with radioactive fallout, and that represented a significant shift away from atmospheric testing. This treaty also recognized that prohibitions on testing severely reduced the chances that new weapon systems would be produced. While many argued in favor of a comprehensive test ban as a way of reducing the arms race, the resulting agreement limited but did not stop nuclear tests (it allowed underground testing). A comprehensive prohibition on testing remains a significant issue today.

The fourth category in Table 5 includes the major SALT agreements, which focus on strategic weapons and the qualitative improvements in delivery systems (such as MIRV). Two points should be noted about these agreements. Provisions of the ABM Treaty are currently under debate as part of the Reagan Strategic Defense Initiative. Previous bilateral agreement to follow the provisions of SALT II, which has not been ratified by the United States, is now in doubt. The United States has argued that the Soviet Union has violated the principles and limitations of SALT II, thereby freeing the United States from conforming to its restrictions. Recently the United States has officially exceeded SALT II limits.

There remains substantial public support for arms control and actual arms reduction (or disarmament). International support for this position is included in the Nonproliferation Treaty, which commits nuclear nations to negotiations to halt the arms race. While arms control talks normally are seen as efforts in this direction, they have not to this point led to actual reductions in the arsenals. They have led to changes in the structure of the arsenals, and to the size of the strategic forces, for example,

Table 5.
Treaties to Control Nuclear Weapons

I. To prevent the spread of nuclear weapons

Antarctic Treaty — December 1, 1959 — 22 states
Bans any military uses of Antarctic and specifically prohibits nuclear tests and waste.

Outer Space Treaty — January 27, 1967 — 76 states
Bans nuclear weapons in earth orbit and their stationing in outer space.

Latin American Nuclear-Free Zone Treaty — February 14, 1967 — 22 states
Bans testing, possession, development of nuclear weapons and requires safeguards on facilities. All Latin American states except Argentina, Brazil, Chile, Cuba are parties to the treaty.

Non-Proliferation Treaty — July 1, 1968 — 115 states
Bans transfer of weapons or weapons technology to non-nuclear-weapons states. Requires safeguards on their facilities. Commits nuclear-weapon states to negotiations to halt the arms race.

Seabed Treaty — February 11, 1971 — 66 states
Bans nuclear weapons on the seabed beyond a 12-mile coastal limit.

II. To reduce the risk of nuclear war between the U.S. and USSR

Hot Line Agreement and Modernization — June 20, 1963 — US–USSR Agreement
Establishes direct radio and wire-telegraph links between Moscow and Washington to ensure communication between heads of government in times of crisis. A second agreement in 1971 provided for satellite communication circuits to improve reliability.

Accidents Measures Agreement — September 30, 1971 — US–USSR
Pledges U.S. and USSR to improve safeguards against accidental or unauthorized use of nuclear weapons.

Prevention of Nuclear War Agreement — June 22, 1973 — US–USSR
Requires consultation between the two countries if there is a danger of nuclear war.

III. To limit nuclear testing

Partial Test Ban Treaty — August 5, 1963 — 108 states
Bans nuclear weapons tests in the atmosphere, outer space, or underwater. Bans underground explosions which cause release of radioactive debris beyond the state's borders.

Threshold Test Ban Treaty — July 3, 1974 — US–USSR
Bans underground tests having a yield above 150 kilotons (150,000 tons of TNT).

Peaceful Nuclear Explosions Treaty — May 28, 1976 — US–USSR
Bans "group explosions" with aggregate yield over 1,500 kilotons and requires on-site observers of group explosions with yield over 150 kilotons.

IV. To limit nuclear weapons

ABM Treaty (SALT) and Protocol — May 26, 1972 — US–USSR
Limits anti-ballistic missile systems to two deployment areas on each side. Subsequently, in Protocol of 1974, each side was restricted to one deployment area.

SALT I Interim Agreement — May 26, 1972 — US–USSR
Freezes the number of strategic ballistic missile launchers, and permits an increase in SLBM launchers up to an agreed level only with equivalent dismantling of older ICBM or SLBM launchers.

SALT II* — June 18, 1979 — US–USSR
Limits numbers of strategic nuclear delivery vehicles, launchers of MIRV'd missiles, bombers with long-range cruise missiles, warheads on existing ICBM's, etc. Bans testing or deploying new ICBM's.

*Not yet ratified.
Source: Louis Rene Beres, *Mimicking Sisyphus: America's Countervailing Nuclear Strategy* (Lexington, Mass.: Heath, 1983), pp. 76–77.

but not to real reductions. Overall, evaluation of the utility of arms control efforts tends to be a function of one's perspective. For the person viewing such negotiations as reflective of the need for disarmament, arms control has failed. New weapons are built, new technologies are adapted to existing systems, and dependence on such systems is increased. At the same time, the concept of negotiated agreements is approved as a political means to control further increases. To the person viewing arms control talks as a means of stabilizing deterrence, they have been a useful and necessary mechanism of tension-reduction. To the person rejecting deterrence as a viable policy, arms control talks are an obstacle to the maintenance of superiority.

Persons wishing to read more in this area will find primary documents in Dupuy and Hammerman, U.S. Arms Control and Disarmament Agency (1961, 1975), and the annual reports prepared by the Stockholm International Peace Research Institute 1972. General analyses of arms control are provided in Barton, Dougherty, and Ranger. Critiques of the arms control process are found in Myrdal, and Platt and Weiler. Readers on the subject include those by Burt (1982), Russet and Blair, and Stockholm International Peace Research Institute (1982). Also useful are two recent journal issues: the winter 1978–79 issue of *International Security*, entitled "Stability and Strategic Arms Control" and the 1981 issue of *International Journal*, entitled "Arms Control" (see section 4 of the appendix for a description of these and other journals).

Military Expenditures

The continuing superpower rivalry between the United States and the Soviet Union and their direct allies is reflected not only in their nuclear arsenals but in overall military expenditures. It has been estimated that global military spending today is thirteen times that of a half century ago, even when controlling on prices, and twice the rate of global gross national product growth.[8] This intensification of spending reflects not only the dominant patterns of the United States and Soviet Union, but also the spread and growth of military institutions in previous colonies and the so-called ratchet effect in which military spending tends to increase dramatically during times of crises and conflict (for example, immediately before and during wars) and to remain relatively high afterward. Post-World War I and II global military spending reverted not to prewar levels but rather to a maximum (mobilization) level. The higher plateau of spending tends to stabilize until further crises occur. Two reasons for this phenomenon are cited. First, the rapid increase in spending for war is viewed as proof that a state can absorb the increased cost; and second,

8. Sivard, *World Military and Social Expenditures 1983*, p. 6.

the development of a large military and supporting industrial base becomes defined as integral to the economic activity of the country.[9]

Distribution. The dominant effect of the superpowers on global spending becomes more evident as the distribution and composition of military spending is examined. Table 6 reports spending during the past decade. Total spending in constant dollars has increased 29 percent. This increase marks important differences in the distribution of spending. First, the two major defense alliances, NATO and WTO (Warsaw Treaty Organization), have remained the dominant spenders since their formation over thirty years ago. Together, these alliances accounted for 75 percent of all global expenditures in 1974; by 1983, their share had decreased slightly to 72 percent. Second, NATO countries increased expenditures by 29 percent during the decade compared to a 16 percent increase among WTO countries. Within the alliances, a different picture emerges. Military spending by the United States declined in the mid-1970s (reflecting the U.S. departure from Southeast Asia) but rose steeply in the early 1980s; its overall increase of 30 percent outstripped the average 24 percent increase among other NATO countries. Military spending by the Soviet Union increased by an estimated 14 percent, compared to a 35 percent increase by the other WTO countries. Third, developing countries, while accounting for only one-fourth the global military budget, increased at a significantly higher rate than that found in NATO or WTO countries. Among developing countries, important geographic differences emerge: the major oil-exporting nations increased their spending by 92 percent (this figure is heavily influenced by wars in the area); South American countries increased their spending by 84 percent, followed by a 60 percent increase among the remaining countries (listed in table 6 as "rest of the world").

Composition. Data on military expenditures for twelve countries that account for about three-quarters of global spending indicate that approximately 30 percent of military costs is associated with uniformed personnel, 30 percent with weapons and equipment, another 30 percent with operations, maintenance, and construction, and only 10 percent with research and development.[10] Two of these categories merit elaboration.

Table 7 compares the changes in military personnel for developed (primarily NATO and WTO) and developing countries during the 1960 to 1981 period. While the number of military personnel has increased by 36 percent, a crucial difference is evident: developing countries have experienced an 80 percent increase in military personnel, while developed countries have experienced a 3 percent decline. By 1981, two of every

9. Stockholm International Peace Research Institute (SIPRI), *Arms Uncontrolled* (Cambridge: Harvard Univ. Pr., 1975, p. 7.
    10. Ibid., p. 12.

Table 6.
World Military Expenditure, 1974–1983 (in constant price figures*)

	1974	1975	1976	1977	1978	1979	1980	1981	1982	1983
United States	$143,656	$139,277	$131,712	$137,126	$137,938	$138,796	$143,981	$153,884	$167,673	$186,544
Other NATO countries	97,606	99,582	101,524	103,214	107,037	109,355	112,297	113,234	116,153	120,627
Total NATO countries	241,262	238,859	233,236	240,340	244,975	248,151	256,278	267,118	283,826	307,171
Soviet Union	(120,700)	(122,600)	(124,200)	(126,100)	(128,000)	(129,600)	(131,500)	(133,700)	(135,500)	(137,600)
Other WTO countries	10,166	10,942	11,418	11,735	12,073	12,228	12,400	12,550	13,135	13,530
Total WTO countries	(130,866)	(133,542)	(135,618)	(137,835)	(140,073)	(141,828)	(143,900)	(146,250)	(148,635)	(151,130)
Other European countries	12,903	13,423	14,047	14,029	14,232	14,979	15,470	15,348	15,291	15,338
Middle East	28,481	35,076	38,670	37,256	37,017	38,893	(40,695)	(45,990)	(52,350)	(50,000)
South Asia	4,569	5,006	5,681	5,497	5,739	6,220	6,460	6,895	7,620	7,865
Far East (excluding China)	(17,970)	(19,930)	(21,750)	(23,220)	(25,630)	(26,610)	(27,600)	(28,790)	(31,100)	(32,950)
China	(35,000)	(36,800)	(37,600)	(36,100)	(40,500)	(52,700)	(42,600)	(36,300)	(37,700)	(35,800)
Oceania	3,976	3,845	3,831	3,848	3,913	4,029	4,270	4,488	4,623	4,868
Africa (excluding Egypt)	9,489	11,416	12,618	12,971	13,198	(13,526)	(13,555)	(13,590)	(13,800)	(14,100)
Central America	1,351	(1,502)	(1,700)	2,173	2,312	2,468	2,484	2,625	2,815	(2,825)
South America	7,998	8,911	9,444	10,170	9,980	9,941	10,230	10,584	(15,745)	(14,745)
World total	$493,865	$508,310	$514,195	$523,539	$537,569	$559,345	$563,542	$577,978	$613,500	$636,790
Industrial market economies	258,406	255,354	249,849	257,175	262,836	267,659	276,931	287,357	304,573	328,944
Nonmarket economies	168,252	173,151	176,341	177,654	184,557	198,868	190,991	187,335	191,486	192,661
Major oil-exporting countries	25,282	32,990	36,962	35,508	37,102	37,807	40,221	45,235	51,510	48,745
Rest of the world	40,817	45,583	49,698	51,774	51,495	53,375	53,701	56,274	64,011	64,408
With 1981 per capita GNP:										
Less than U.S. $410	6,878	7,225	7,756	7,379	8,004	8,458	8,646	9,132	9,964	10,207
U.S. $410–1,699	11,437	13,792	14,859	15,778	14,064	14,835	13,963	14,781	15,777	16,641
Greater than U.S. $1,700	22,502	24,566	27,083	28,617	29,427	30,082	31,092	32,361	38,270	37,560

*Figures are in millions of U.S. dollars at 1980 prices and exchange rates. Totals may not add up due to rounding.
Source: *SIPRI Yearbook 1984*, p. 117.

three members of the armed forces were in developing countries. The increase in military personnel translates into higher per capita military expenditures: during the 1960 to 1981 period, developing countries experienced a sixfold increase in per capita expenditures, compared to a threefold increase among developed countries.[11] Among developed nations, the greater cost for personnel follows a decline in conscription systems and concomitant demand for more highly paid skilled military personnel.[12]

While expenditures for research and development (primarily weapons) account for only about 10 percent of global military expenditures, such spending is highly concentrated with over 90 percent in six countries (five of which have nuclear weapon programs): United States, Soviet Union, United Kingdom, China, France, and West Germany. The United States and the Soviet Union alone account for approximately 75 percent of this total.[13]

The research and development effort in the United States consumes a large share of scientific personnel. While research and development account for only about 12 percent of military budget, they represent 64 percent of all government-sponsored research and development expenditures and 28 percent for the country as a whole.[14] The majority of the military research and development expenditures is oriented toward more sophisticated weapons rather than increasing existing arsenals. As a consequence, advances in weapon technology are exported throughout the world. Each upgrade facilitates dependence on those countries that dominate the research and development process and makes available outdated but highly lethal weapons to a larger group of nations (as developed nations dump older weapons in favor of new ones).

In this section we have emphasized the continuing dominance of the Soviet Union and the United States on global military spending. We indicated that while military spending is increasing on a global scale, changes in the distribution and composition of such spending reflect a direct relationship to the nuclear arms race. The spending patterns of developed nuclear nations are characterized by a reduced dependence on armed personnel and an increased emphasis on weapons—that is, weapons, particularly nuclear weapons, compensate for reduced standing military organizations. Developing countries have come to rely on increasing numbers of military personnel, and as we see in the next section, on the weapons developed by the superpowers. There is a dominant regime—a hierarchical military system in which those at the top not only

11. Sivard, *World Military and Social Expenditures 1983*, p. 32.
12. SIPRI, *Arms Uncontrolled*, pp. 14–15.
13. Stockholm International Peace Research Institute (SIPRI), *World Armament and Disarmament: SIPRI, Yearbook 1984* (Philadelphia: Taylor & Francis, 1984), p. 165.
14. Ibid., pp. 170–71.

Table 7.
Changes in Military Personnel

	1960		1965		1970		1975		1981	
	Number	*Percent*	*Number*	*Percent*	*Number*	*Percent*	*Number*	*Percent*	*Number*	*Percent*
Armed forces (in thousands)										
Total world	18,550	100	19,528	100	21,484	100	21,863	100	25,311	100
Developed countries	9,851	53	9,711	50	10,146	47	9,550	43	9,581	37
Developing countries	8,699	47	9,817	50	11,338	53	12,313	57	15,730	63

Source: Ruth Leger Sivard, *World Military and Social Expenditures, 1983* (Washington, D.C.: World Priorities, 1983), p.32.

control the majority of the world's nuclear arsenals but shape the direction and scope of conventional military expansion throughout the world.

Such a situation is analogous to colonialism—the core countries, while competing among themselves, also dominate the rest of the world. For annual and long-term analyses of expenditures, see the reports issued by the International Institute for Strategic Studies, Stockholm International Peace Research Institute, the Arms Control and Disarmament Agency, and Sivard.

Arms Trade and Transfer

The emergent global arms trade played a crucial role in the global militarization of the past half century. A relatively small number of countries dominate arms exports, although this number has recently begun to expand.[15] The United States and the Soviet Union account for about 70 percent of total arms exports. During the past five years, the Soviet share of global arms exports has declined from 46 percent to 30 percent, while the U.S. share has increased from 26 percent to 39 percent. During this period of time, about 70 percent of Soviet arms sales have been to developing countries, making the Soviet Union their chief supplier. American arms exports are evenly divided between industrialized and developing countries.

During the past forty years, almost all wars have occurred in the developing countries. These wars, whether civil or interstate in character, have been heavily dependent upon arms supplied by the industrialized nations. This may be seen in the changing geographical distribution of arms importers during the past twenty years: Today almost one-half of arms exports go to the Middle East, 20 percent to Africa, and 13 percent to Latin America. Significant reductions in shares of total imports have been experienced by countries in the Far East (a reduction from 31 percent to 11 percent) and South Asia (a drop from 12.5 percent to 7.7 percent). Over the past five years, the volume of Third World imports has stagnated. This reflects less a lowered dependency on imports than an inability to purchase caused by worsening national debt and economic crises.

American and Soviet exports to the Third World remain an important element in each nation's foreign policy objectives. The general pattern for the Soviet Union has been to concentrate its arms exports on a smaller number of countries, while the United States spreads a lower volume among a larger number of countries.[16] The other major arms exporters, particularly France and West Germany, tend to sell arms from economic motivation; the United States and the Soviet Union export more for polit-

15. This section borrows heavily from SIPRI, *World Armament and Disarmament: SIPRI Yearbook 1984*, pp. 175–210.
16. Ibid., pp. 176–79.

ical purposes. Such a distribution was recently called into question as both the United States and Soviet Union exported weapons to each side of the Iran-Iraq war.[17] The arms trade with other industrialized nations is characterized by mutual trade and specialization.[18] Many developed societies find it advantageous to import certain arms while exporting others, thereby reducing the need to establish all-inclusive arms industries within their own political boundaries. This, when combined with the recent increase in the number of exporting countries, has led to an overall intensification of competition for exports to both developing and developed countries.[19]

The literature on arms trade and transfer is quite large. For basic sources, see the reports issued by the Arms Control and Disarmament Agency and the annual reports issued by SIPRI. Popular discussions can be found in Sampson and Klare. Kaldor and Eide; Blechman, Nolan, and Platt; Burt; Farley, Kaplan, and Lewis; Frank; Harkavy; and Newman and Harkavy should be consulted for more detailed analyses.

Impact on Conventional Weaponry

The primary change in conventional weaponry of the recent past has involved increasing qualitative sophistication. A major part of this change has resulted from advances in command, control, communications, and intelligence systems, major parts of which are applicable to both nuclear and nonnuclear weapons:

> A prime example of this is the integrated chip which makes possible a large amount of information to be incorporated into various weapon systems: navigation is increased significantly in missiles and munitions; . . . advanced infrared sensors make weapons more effective; . . . innovations in design have led to the "submunitions revolution" resulting in the spread of destructive energy over a larger area.[20]

The increasing sophistication of conventional weaponry has resulted in increased costs per weapon, a greater likelihood of cost overruns as a new technology comes to be adapted in the period between initial research and subsequent development, and increased maintenance costs associated with the high level of sophistication (for example, electronic complexity) of the weapon themselves.

A second trend is toward increasing production by developed countries of conventional weapons for mass destruction—both bombs and weapons in the chemical and biological warfare category. Development in binary weapons has increased the longevity and stock of the weapons:

---

17. Ibid., pp. 195–201.
18. SIPRI, *Arms Uncontrolled*, p. 36.
19. SIPRI, *World Armament and Disarmament: SIPRI Yearbook 1984*, p. 175.
20. Craig and Jungerman, *Nuclear Arms Race*, pp. 103–4.

The Reagan administration has proposed building a factory for binary nerve gas munitions. The capacity would be for 20,000 rounds of 155 millimeter shells per month. Five hundred-pound binary spray bombs for aerial use would also be available . . . . Binary weapons are being prepared for use in a wide range of platforms, including ground-launched cruise missiles.[21]

As existing arms become more sophisticated and more lethal, and as production of conventional weapons of mass destruction increase, the distinctions between nuclear and conventional weapons becomes blurred.

The reader interested in changing technology in weapon systems should consult Barnaby and Huisken, Klare, Carlton and Schaerf, Kaldor (1981), Cane, Feld, and Fallows.

Horizontal Nuclear Proliferation

The continued development of nuclear weapons by the United States and Soviet Union is often termed the vertical arms race, emphasizing their hegemonic position with possession of 95 percent of the world's nuclear arms. In contrast, the term horizontal arms race is used to describe the spread of nuclear weapons to other countries. This spread is viewed as a significant problem insofar as (1) it is economically wasteful (for example, costs outweigh potential benefits); (2) it is politically unproductive (for example, it does not fulfill foreign policy objectives or substantially aid in defense or security), and it threatens international stability; (3) it facilitates a dependence on nuclear rather than conventional weaponry (thereby increasing the consequences of conflict); and (4) it increases the likelihood of war (accidental war, terrorist attack, or intentional state action). These criteria tend to lead to fundamentally different interpretations of the problem of horizontal and vertical proliferation. That is, while the United States and the Soviet Union may view continued proliferation of their own arsenals as stabilizing and disarmament as destabilizing, similar lines of development by nonnuclear nations are normally viewed as dangerous. Nonnuclear countries may view the acquisition of a nuclear capacity as the shortest and most efficient means of gaining power over rivals and in reducing the monopolistic position of the United States and the Soviet Union.

Over the past three decades, there has been general, though not universal, agreement that the spread of nuclear arms to an increasing number of nations was politically destabilizing, economically prohibitive, and militarily dangerous. Concern over horizontal proliferation increased significantly as knowledge became available on the link between nuclear power and nuclear weapons. Beginning with the establishment of the International Atomic Energy Agency (IAEA) in 1957, there has developed what has been termed the nonproliferation regime:

21. Ibid., pp. 105–6.

A loose structure of treaty commitments verified by international inspection, not to acquire nuclear weapons; informal and voluntary understandings of nuclear supplier states to limit certain nuclear exports, to acquire safeguards for others, and to limit their nuclear cooperation to the least dangerous technologies; bilateral agreements between some nuclear supplier states and their clients; and a general disposition against nuclear weapons.[22]

Multilateral nonproliferation treaties pledging not to test or develop nuclear weapons include the Partial Test Ban Treaty (1963), which prohibits nuclear explosions in the atmosphere or underwater; the Treaty for the Prohibition of Nuclear Weapons in Latin America; and the Treaty on the Non-Proliferation of Nuclear Weapons (NPT), which prohibits the transfer by nuclear states of nuclear weapons to nonnuclear states. The latter two treaties assign verification responsibility to the IAEA, particularly in ensuring detection of fuel-cycle diversions. The success of this regime depends on whether one defines proliferation as actual testing and production or as the capacity to make nuclear weapons rapidly. The data appear to indicate success in limiting production, but also show that a number of states are closer to the capacity to make weapons (since the passage of the NPT.)[23]

Table 8 outlines the major pressures, pro and con, on the regime today. These pressures indicate its potential volatility, particularly the continued necessity for suppliers to maintain constraint, to improve safeguards, and to expand the number of signatories. On the other hand, technological developments, particularly in breeding technology, coupled with the potential attraction of nuclear energy and weapons for countries approaching the necessary levels of capacity and motivation, suggest that one might well expect further proliferation in the future. Argentina, the Federal Republic of Germany, India, and Israel are likely candidates.

For analyses of the problems of horizontal proliferation and the effectiveness of existing agreements, see Brenner, Dunn, Meyer (1983), Epstein (1976), Quester (1981) and Jones. A useful review essay is provided by Quester (1983), and a annotated bibliography by Graham and Evers (1978).

Summary

The global trend in militarization since World War II has been dominated by the superpower arms race between the United States and the Soviet Union. As the major spenders, weapon developers, and exporters, they strongly shape the international structure of military orga-

---

22. Warren Donnelly, "Changed Pressures on the Non-Proliferation Regime," in SIPRI, *World Armament and Disarmament: SIPRI Yearbook 1983*, p. 69.
23. Ibid., p. 74.

Table 8.
Horizontal Nonproliferation Regime

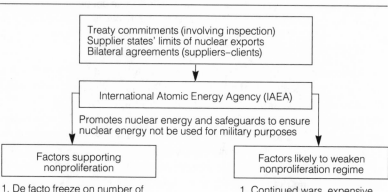

Factors supporting nonproliferation	Factors likely to weaken nonproliferation regime
1. De facto freeze on number of nuclear weapon states	1. Continued wars, expensive conventional weapons
2. No terrorist use to date	2. U.S.–USSR tension
3. Slowdown in nuclear power	3. Doubts about security assurances
4. Weakness in world nuclear industries	4. Hold-outs from no-nuclear weapons pledges (such as Brazil, Argentina, Israel, Pakistan)
5. Nuclear difficulties of threshold states (e.g., Brazil, Argentina)	
6. Diminished use of highly enriched uranium	5. No progress in nuclear disarmament
7. Supplier restraint	6. Development of nuclear power technologies (such as plutonium breeding)
8. Predisposition against nuclear weapons	
9. Improvements in safeguard technologies	7. Spread of sensitive (enrichment) technology
	8. Challenges to verification of no-weapons pledges
	9. Existence of unsafeguarded facilities
	10. Politicization of IAEA
	11. Changing definition of proliferation
	12. Agitation against supply restrictions

Source: Adapted from Warren Donnelly, "Changing Pressures on the Non-Proliferation Regime," *Armaments and Disarmaments. SIPRI Yearbook, 1983* (London: Taylor & Francis, 1983), pp. 69–95.

nization. They have stockpiled tens of thousands of nuclear weapons in addition to having large standing military organizations, defense industries, and conventional arsenals. Periodically they increase the size or efficiency of their nuclear arsenals, widening the gap between their military status and that of the rest of the world. They compete for political allies and markets for their weapons. They have acted to block others from building nuclear arsenals while facilitating proliferation of conventional armaments. They have agreed at various points to restrict development of certain weapons while allowing quantitative improvements in existing systems and open development of others. The rivalry continues to build, while involving more of the world in its consequences.

Socioeconomic and Political Consequences of the Arms Race

In recent years, there has been mounting criticism that the costs associated with the arms race have threatened social, economic, and political development in industrialized countries and developing countries. The basic thrust of these arguments is that the pressure for increased military spending and weapon acquisition diverts resources away from more productive or necessary activities, such as producer and consumer goods and services, and results in a lower standard of living, reduced productivity and growth, and so on. Such a perspective does not question the idea of providing for national security, but argues that recent military expenditures, in size and scope, do not provide such security. Two dimensions are described in this section: (1) the economic costs of the arms race to industrial societies; and (2) the economic, military, and political consequences of the arms race on developing countries.

Economic Consequences for Industrial Societies

In industrial societies today one finds the highest rate of military spending, large standing military organizations, an industrial base oriented toward military production in varying degrees, and enormous conventional and, in certain cases, nuclear arsenals. In earlier times, mobilization for war occurred quickly and temporarily. Industrial realignment had severe but short-run effects on the economy. Since World War II, many of the industrial countries have developed large, permanent defense industries, weapon programs, and standing military organizations as peacetime necessities.

Table 9 illustrates this point, reporting the military burden on selected industrialized nations in terms of its relative size of government spending or as a percentage of the gross national product. The first column, the military part of the state budget, indicates a wide range of 6 percent to 25 percent; the countries with the two highest proportions of military spending, the United States and Switzerland, have fundamentally different military organizations. Switzerland focuses the vast majority of its resources on defending its borders and maintaining a well-trained militia. The United States expends its resources on external targets, with almost no money directed toward literal protection of its borders. The two largest categories of the U.S. budget, general purpose forces and its nuclear arsenal, are oriented toward warfighting elsewhere, especially Europe.

Table 9.
Military Burden for Selected Countries, 1981

Country	Military Expenditures as a Percentage of Government Expenditures*	Military Expenditures as a Percentage of Gross National Product†
Australia	11.9	2.5
Canada	8.7	1.9
France	18.1	4.0
West Germany	22.3	3.2
Japan	5.9	.9
Spain	11.3	1.9
Switzerland	25.0	2.0
United Kingdom	14.3	5.2
United States	24.7	5.6

*Ruth Sivard, *World Military and Social Expenditures 1983* (Washington, D.C.: World Priorities, 1983), pp. 33–34.
†Stockholm International Peace Research Institute, *World Armaments and Disarmament. SIPRI Yearbook 1984* (London: Taylor & Francis, 1984), p. 80.

The lowest relative military burden is found in Japan and Canada. The low emphasis in Japan is a direct consequence of Western pressures after World War II, although there is now pressure on Japan to expand its military emphasis. Canada, similar to the United States in not facing conventional threats to its homeland, maintains a relatively small standing military organization.

The second column in table 9 reflects the proportion of each country's economic activity that is diverted to military ends. Again, the range is large, with the United States directing significantly more of its economic activity than every other country with the exception of the United Kingdom. These figures for the United States would be much higher today as a result of the buildup during the past five years, when budget authority and actual spending levels increased more than 40 percent after inflation.

While drawing on large amounts of natural and social resources, these emerging military complexes were viewed as good for the economy. This view is increasingly disputed along a series of economic decisions; the thrust of these arguments is that *long-term* diversion of resources for unnecessary (for example, redundant) armament programs leads to economic decay, primarily in the form of higher inflation, high unemployment, and lessened productivity. Most of the discussion below focuses on the United States; the arguments, however, are seen to be valid for industrial capitalist societies in general. The effect on centrally planned industrial economies is somewhat different. (See Dumas for a discussion of this difference.)

Drain of Key Resources.   One element in the criticism of the arms race deals with scarce resources that are diverted to military purposes.

1. Financial resources diverted from social or infrastructure programs lead to a long-term neglect of civilian-oriented equipment and facilities (such as railroads and industrial equipment) and civilian-oriented technological development.[24] Financial resources are also diverted to repay interest on long-term government debt, a major part of which is associated with previous wars.

2. Competition intensifies between military and private producers for vast amounts on nonrenewable raw materials, thereby raising the costs for such materials and depleting supplies.

3. Research and development capability is severely diminished. For example, it is estimated that 20 percent of the technical and scientific talent of the United States is involved in military research and development. This lowers the rate of private technological development. And while it is often argued that military research has a variety of spin-off benefits to the civilian economy, recent studies question the impact of such spin-offs.[25]

Heightened Inflation. The arms race results in continual inflationary pressures because of the economic nature of military goods such as weapons and military equipment. Such goods do not produce economic value in the traditional sense of contributing to the material standard of living. Since the outcomes of the expenditures for military goods lead to a subsequent increase in demand for producer and consumer goods without a corresponding increase in supply of such goods, heightened inflation results.[26]

Reduced Productivity Growth and International Competitiveness. A further consequence of the military burden on the economic system is reduced growth in productivity. Evidence supporting this view is presented in Figure 7, which compares the average level of military spending as a share of gross national product with increases in manufacturing productivity (measured by growth in output per hour) for selected industrial countries during the past twenty years. A strong negative correlation exists: the higher the military burden, the lower the degree of growth in productivity. To the extent that productivity growth is a crucial element in the ability to compete in international markets, the result is often a negative balance of trade.

Employment. Military expenditures, particularly those involved in the production of weapon systems, have traditionally been argued to be beneficial to the economy because of the possibility that jobs will be created. Recently, this argument has been attacked on two grounds. First, recent studies indicate that military spending creates fewer job opportu-

24. Lloyd J. Dumas, *The Political Economy of Arms Reduction: Reversing Economic Decay* (Boulder, Colo.: Westview, 1982), p. 2.
25. Ibid., pp. 11–20.
26. Ibid., pp. 5–6.

Figure 7.

Military Spending vs. Manufacturing Productivity, Average Percent, 1960–1979.

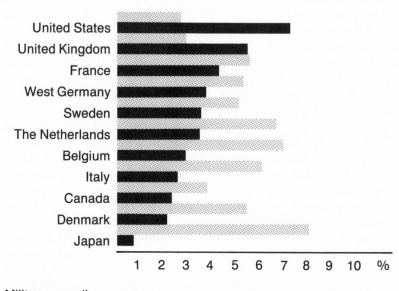

United States
United Kingdom
France
West Germany
Sweden
The Netherlands
Belgium
Italy
Canada
Denmark
Japan

1   2   3   4   5   6   7   8   9   10   %

Military spending:
share of gross domestic product

Productivity:
growth in output per hour

Source: Council on Economic Priorities. The Costs and Consequences of Reagan's Military Building," 1982.

nities than the same expenditures for nondefense (for example, social) sectors.[27] Other studies have found that the job creation potential of defense expenditures does not affect the unemployed; instead, demand increases for skilled workers, such as scientists and engineers.[28] This demand lowers the number of skilled workers available for nondefense related work. To explore the effects of the arms race on developed societies, see Adams; Dellums, Miller, and Halterman; Dumas, Melman (1970, 1971), United Nations (1972, 1977), Boston Study Group, Gansler, Klare (1976), Long and Ruppy, Stevenson, and Yarmolinsky.

27. Kenneth A. Bertsch and Linda Shaw, *The Nuclear Weapons Industry* (Washington, D.C.: Investor Responsibility Research Center, 1984), p. 88.
28. Ibid., pp. 88–89.

Effects on Developing Nations

Debates on the effects of the arms race have mostly centered on East-West relations (the rivalry between capitalist and centrally planned societies). Analysts have recently begun to emphasize the detrimental effects on developing societies. Basically, these analyses have taken a Third World perspective, viewing developing nations as dependent economically, militarily, and politically on the core industrial powers (capitalist and socialist). Competition within the core for military resources and dominance results in lessened development in the periphery and an increasing gap in the global distribution of wealth.

Such studies emphasize the idea that a global military order has arisen since World War II, characterized by increased trade and transfer of armaments, dependence on a few countries for weapon innovation and development, and the unequal relationship between the developed and developing world—one in which the developed world exploits the developing countries through consumption of their raw materials, misdirection of their economic resources, and support for despotic governments. Three general consequences are delineated in this perspective.

Economic Consequences. Developing countries are generally characterized by nondiversified economies, a relatively low-skilled labor force, negative balance of trade, limited state revenues, high unemployment, low rates of investment capital, and structures of production insufficient to meet the basic needs of its population.[29] In this context, the arms race, manifest in conventional arms trade and the structure of foreign aid, drains both public and private consumption and constrains growth. First, the volume and the structure of investment, a key component in economic growth, is negatively affected: investment needed to reduce problems in housing, agriculture, health, education, and energy is diverted to military purposes. Second, the size and composition of the workforce is effected in two ways: increased military spending has not reduced unemployment in ways that other spending might (for example, demand stimulus); skilled personnel, a scarce commodity, are likely to be involved in direct and indirect support for the military, thereby diverting such labor from work on significant social needs. Third, the rate of technological change is lowered in two ways. Similar to the effects in industrial societies, technological change oriented toward social and infrastructure needs would have greater benefits to people in developing societies; and the purported spin-off benefits of military research are actually quite low given the degree to which research is most generally oriented toward the qualitative performance characteristics of weapons—activities that have little, if any, civilian transferability.

29. This section draws heavily from U.N. Department of Political and Security Council Affairs, *Economic and Social Consequences of the Arms Race and of Military Expenditures* (New York: United Nations, 1978), pp. 39–57.

Effects on the Military.   The developing world's share of global military expenditures remains a minor fraction; at the same time, its share of global expenditures has doubled in the past two decades. The dynamics of the arms race, both conventional and nuclear, are such that there is an inherent push to adapt not necessarily the most lethal armaments but the most up-to-date ones. The more traditional pattern of receiving obsolete weapons from the developed world has been replaced by a tendency to import sophisticated new weapons. This is done to maintain an advantage over potential opponents and in some cases as a cost-effective response to diminishing state revenues—that is, it may be more cost effective to purchase sophisticated weaponry to compensate for inadequate military organizations. Exacerbating this is the trend of military aid and foreign aid from the industrialized nations to reflect their own political considerations rather than the security needs of the developing nation.

Effects on Political Institutions.   A third area of concern deals with the effect of the arms race on the internal political institutions of the developing nations and the more general impact on international political relations. In the first case, the focus is on both the pervasive influence of military rule and the role of the military in shaping the development process. Military dominance of political institutions is a widespread, though not universal, pattern. The role of the military in many societies becomes twofold—defense against external threat and internal repression:

> Here governments can get trapped in an impossible situation where an increasing burden of military expenditures further delays economic and social progress, freezes social structures and exacerbates social tension, while other policies seem to be precluded by the context of confrontation and arms race with neighboring countries. The conjunction of external and domestic confrontation, both of them temporarily stabilized through military build-up but ultimately exacerbated by it, can give rise to a particularly precarious situation.[30]

In terms of international relations, the arms race is seen as facilitating military power over economic cooperation, with the economic and social issues becoming secondary to military consideration:

> The arms race represents a waste of resources, a diversion of the economy away from its humanitarian purposes, a hinderance to national development efforts and the threat to democratic processes. But its most important feature is that in effect it undermines national, regional, and international security. It involves the constant risk of war engaging the largest Powers, including nuclear war, and it is accompanied by an endless series of wars at lower levels. It raises an even greater barrier against the atmosphere in which the role of force in international relations may be downgraded. In addition, it impedes relations between countries, affecting the volume and di-

30. Ibid., p. 56.

rection of exchanges, diminishing the role of cooperation among states and obstructing efforts toward establishing a new international economic order on a more equitable basis.[31]

Persons interested in the effects of the arms race on developing countries should consult Eide and Thee, United Nations (1972, 1977), and UNESCO (1978) for general treatments. Albrecht (1974, 1975), International Peace Research Association (IPRA), Jolly, Lumsden, Oeberg, Stockholm International Peace Research Institute (1971), Lock and Wulf, and Benoit (1963, 1973) examine the impact of the arms race on economic development. Political consequences are covered in Blechman (1976, 1982), Doorn, Eide, Fidel, Frank, Gutteridge (1962, 1964), Kaldor and Eide (1979), Kemp, Klare and Arnson, and Varynynen (1973, 1980).

This chapter discusses many of the ongoing costs and consequences of the arms race. Even in the absence of nuclear war, these costs have increasingly become matters of national and international debates. Legitimate questions are being asked that may well require policy changes: Is it necessary to continue building nuclear arsenals? Are arms control negotiations realistically attempting to reduce the nuclear threat, or are they mechanisms to sustain such arsenals? Are the global patterns of weapon trade and transfer aiding or detracting from the developmental process? Will the current nuclear monopoly continue or will other countries seek to gain political and military power through nuclear arsenals? Is the existing military burden on industrialized countries leading to long-term economic decay? What impact does the arms race have on international political relations?

Bibliography

The following bibliography includes references cited in the text and related suggested readings; references are organized according to the major divisions of the chapter. The full bibliographic reference for all sources cited in the text of this chapter may be found in either the annotated or unannotated bibliography corresponding to the division of the text in which the source is cited. Sources that are annotated are the most essential or significant.

Global Militarization

Barnaby, Frank and Ronald Huisken. *Arms Uncontrolled.* Cambridge, Mass.: Harvard Univ. Pr., 1975.

This book is a useful introduction to the conventional and nuclear arms race. The first section describes the dispersion of arms throughout the

31. Ibid., p. 31.

world, focusing on changing military organization as nations come to depend on armaments. The next section examines "tactical" weapons systems, including chemical and biological weapons, and antisubmarine technology. The final two sections look at nuclear arms and attempts at arms control. The book contains a good deal of useful data without being overly technical.

Bottome, Edgar. *The Balance of Terror: Nuclear Weapons and the Illusion of Security 1945–1985*. Boston: Beacon, 1986.

This is a revised and updated version of the 1971 classic critique of the arms race. The basic argument is that the balance of power concept in international politics was replaced in the 1950s by a balance of terror (a nation could no longer claim to be able to protect or defend its population except through threatening another's population) and greater insecurity. The examination of the arms race is structured around three questions: (1) what purpose do nuclear weapons allegedly serve in the conduct of a nation's foreign policy? (2) under what conditions will these weapons be used? and (3) how did we arrive at the current impasse of massive nuclear overkill? Of particular utility is the discussion of the various alleged "gaps" that have justified American development of weapon systems and the degree to which both the United States and Soviet Union are imperialist in theory and practice. This is a useful and readable critique of nuclear policies and weapons, a resource oriented more for the interested layperson than the specialist or strategist.

Carlton, David and Carlo Schaerf, eds. *Arms Control and Technological Innovation*. New York: Halsted Pr., 1976.

This reader consists of twenty papers presented to the Sixth Course of the International School on Disarmament and Research on Conflicts in 1976. The focus is on the causes, consequences, and control of the arms race. Discussion on the causes of the arms race (conventional and nuclear) stresses international political, military, economic, and domestic political factors. The general argument presented is that the superpower rivalry between the United States and the Soviet Union dominates and propels the arms race horizontally and vertically. The readings on the consequences of the arms race include descriptions and analyses of new weapons and their effects on military strategy and national security policy, arms trade and transfer, and the debate over the benefits and costs of military spending to the economy. A final set of readings explore various dimensions of arms control: bilateral vs. multilateral, comprehensive vs. incremental, the role of international law and so on. The general tone of the book may be characterized as critical of existing structures and policies and supportive of international disarmament endeavors.

Cochran, Thomas B., William Arkin, and Milton Hoenig. *Nuclear Weapons Databook*. Vol. I., *Nuclear Forces and Capabilities*. Cambridge, Mass.: Ballinger, 1984.

This is the first of a planned eight-volume series on the production and deployment of nuclear weapons. The focus of this volume is on the nuclearization of American forces—the dispersion of strategic forces, cruise missiles, nuclear capable aircraft, and so on throughout the armed services.

Three chapters provide background information on warheads, missiles, bombs, and research programs. Subsequent chapters provide detailed information on warheads (for example, function, modifications in design and performance over time, yield, level of deployment), missiles (such as contractors, technical specifications, number of warheads, and cost), and so on. The data are presented in such a way that one may trace the chronological development of major weapon systems. Overall, this work gives the general reader extensive technical material in a straightforward and readable format.

Dunn, Lewis A. *Controlling the Bomb: Nuclear Proliferation in the 1980s*. A Twentieth Century Fund Report. New Haven, Conn.: Yale Univ. Pr., 1982.

This book represents one of many recent examinations of horizontal proliferation. The argument presented supports maintaining nuclear dominance by the superpowers. The spread of nuclear weapons to nonnuclear nations is taken for granted, although many suggestions are made to slow its speed and scope. Rapid horizontal proliferation is viewed as increasing the possibility of war and negatively affecting the stability of the superpower arms race. A major theme of this work is the proactive role America should play in slowing proliferation, for example, through increased conventional support of weapons and troops and threats of sanctions and military force. The position taken here is that it is more plausible to affect the political motivations to acquire nuclear weapons than to restrict the technical knowledge and materials necessary to produce such weapons.

Gerson, Joseph, ed. *The Deadly Connection: Nuclear War and U.S. Intervention*. Cambridge, Mass.: American Friends Service Committee, 1983.

This short book consists of papers delivered at an AFSC conference in 1982. Its intent is to explore the relationship between the arms race and intervention in the Third World. The basic argument presented is that nuclear policies reflect a "war as a political mechanism" orientation. Specifically, it is argued that nuclear policies and weapons facilitate political power moves as the United States attempts to gain advantage in the Third World while denying Soviet advantage. The deadly connection posits direct and indirect relations between conventional and nuclear war as subsets of political rivalry. Papers examining this connection discuss Korea, Vietnam, Central America, South Africa, and so on. While the papers are short, they do suggest that the emphasis on weapons as deterring war ignores the specific ways such weapons facilitate conventional war.

International Institute for Strategic Studies. *The Military Balance, 1984–1985*. Oxford, England: Alden Pr., 1984.

This is another of the annual independent examinations of global military strength and expenditures. Primarily quantitative, the report provides data on size and composition of forces, distribution of weapon systems, in-depth comparisons of Soviet and American and NATO and WTO arsenals and force structures. For the person interested in trend research, this resource is a valuable asset.

Kaldor, Mary. *The Baroque Arsenal*. New York: Hill & Wang, 1981.

The author argues that the peacetime military-industrial complex has propelled a primarily quantitative conventional and nuclear arms race characterized by the development of sophisticated lethal weapons that are incapable of meeting limited military goals (and are therefore counterproductive). Rather than stimulating civilian technology, this arms race has a negative effect on the economy and the utility of the military armaments themselves. The continued development of decadent or baroque armaments results in lower civilian productivity, a distortion of the structure of innovation, and long-term economic decay. Topics covered include the military-industrial complex, the Soviet system of weapon development, the impact of the arms race on developing countries, and the long-term consequences of the arms race on the military and economic institutions of society.

Klare, Michael. *American Arms Supermarket.* Austin: Univ. of Texas Pr., 1984.

This book examines American policy and practices of arms trade and transfer during the post-World War II period. The book looks at the pressures for arms exports and the competing perspectives dominant in policy circles (restrainers vs. supporters). Individual chapters include case studies of Latin America and Iran, arms export policy and implementation, technology transfer, U.S. and Soviet exports to the Third World, and suggestions for an alternate policy framework on arms trade. The book is generally critical of existing American policy and practice, arguing that it does not enhance national security, other foreign policy objectives, or economic viability. This work is useful both in its description of the various dimensions of arms trade and in its analysis of alternative policy directions.

Lifton, Robert Jay and Richard Falk. *Indefensible Weapons: The Political and Psychological Case against Nuclearism.* New York: Basic Books, 1982.

Psychological and political consequences of the arms race are explored in this work through the concept of nuclearism—dependence on nuclear weapons as a means of solving problems (for example, security). In the first half of the book, Lifton, a psychiatrist, examines psychological attempts to cope with nuclear weapons. His treatment includes identification of illusions, deceptions, and individual and collective syndromes that act as barriers to effective action such as a sense of futurelessness or psychic numbing). Drawing on his research on the survivors of Hiroshima, Lifton explores various means of overcoming psychic numbing.

The second half of the book focuses on political dimensions of the arms race. Building on the psychological effects of nuclearism, Falk, a political scientist, examines the degree to which nuclear dependence lessens or threatens the democratic process, the mythology of superiority and inferiority in arms racing and in policies of national security, the political context in which nuclear decisions are made (such as the decision to bomb Hiroshima and Nagasaki), the use of anti-Sovietism as a rationale for American nuclear development, the tendency toward political passivity about nuclear weapons, and suggestions for public action and policy change. This is a useful book in its examination of barriers—political and psychological—to lowered dependence on nuclear weapons and in its articulation of constructive alternative to the arms race.

Pierre, Andrew J. *The Global Politics of Arms Sales.* Princeton, N.J.: Princeton Univ. Pr., 1982.

This book examines various motivations often given for the rapid increase in global arms trade, such as the desire for influence, the need for security, and economic benefits. Each is presented as facilitating the arms race, particularly given the recent increase in Third World nationalism and superpower rivalry between the United States and Soviet Union. Exporters and importers are seen to have reciprocal incentives in maintaining and increasing the arms race even though such incentives may become counterproductive over time. The generally accepted rule of thumb that the superpowers trade arms for ideological reasons while other major exporters seek primarily economic benefits is questioned. Substantial focus is given to the domestic and regional political effects of arms trade on selected Third World countries. Recommendations are made to control the number and type of arms traded to the developing world.

Sivard, Ruth Leger. *World Military and Social Expenditures 1985.* Washington, D.C.: World Priorities, 1985.

This is the tenth in a series of annual reports on national and global trends in military and social spending, conventional wars, global militarization and the arms race (nuclear and conventional). Particular focus is given to the dominant role the U.S. and Soviet rivalry plays in global trends in armament, wars, and development. Data are provided on military expenditures, economic indices, nonmilitary spending, indices of social development (for example, education, health, and nutrition). The data include rank orderings as well as aggregate figures (such as by region, military alliance, or level of development). The interpretation of the data is fundamentally different than that of the U.S. Arms Control and Disarmament Agency, which the author at one time prepared (see below). She argues that the arms race threatens national and international security and diverts resources necessary for social development. The superpower rivalry, manifest most clearly in nuclear arsenals, is viewed as resulting in overkill in unnecessary and costly expansion of nuclear arsenals beyond any legitimate security need. This resource is particularly useful to the person interested in researching global military trends in spending, nuclear arsenals, and the relation between military burden and social development.

Tobias, Sheila, Peter Goudinoff, Stefan Leader, and Shelah Leader. *What Kinds of Guns Are They Buying for Your Butter? A Beginner's Guide to Defense, Weaponry, and Military Spending.* New York: Morrow, 1982.

Written for a popular audience, this work attempts to delineate types of weapons we use, the theory of deterrence, and the process by which weapons are procured. Nine current military or weapons-related controversies are explored. The strength of the book is its ability to "demystify" military related issues.

U.S. Arms Control and Disarmament Agency. *World Military Expenditures and Arms Transfers 1985.* Washington, D.C.: U.S. Arms Control and Disarmament Agency, 1985.

This is the sixteenth in the annual publications of the Arms Control and Disarmament Agency. Yearly data on military spending, numbers in the armed forces, arms imports and exports, GNP, central government spending, dollar value of arms deals, and so on are provided for most countries of the world as well as for categories of countries (for example, regions, military alliances, or level of development). Previous editions included data on nonmilitary social indicators (such as numbers of teachers or doctors) as a comparative reference point. Such data have been dropped, replaced by a strong emphasis on anti-Sovietism. The military expenditures data often differ from independent estimates (see Sivard, International Institute for Social Studies, Stockholm International Peace Research Institute), and recent interpretations appear more ideological. Given these changes, it remains a useful source of data on trends, particularly in the area of arms trade and transfers.

Ball, Desmond. *Politics and Force Levels: The Strategic Missile Program of the Kennedy Administration.* Berkeley, Calif.: University of California Pr., 1981.

Barnaby, Charles. *Prospects for Peace.* New York: Pergamon, 1980.

Barton, John H. *The Politics of Peace: An Evaluation of Arms Control.* Stanford, Calif.: Stanford Univ. Pr., 1981.

_____ and Imai Ryukichi. *Arms Control II: A New Approach to International Security.* Stockholm: Oelgeschlager, 1981.

Baugh, William H. "Major Powers and Their Weak Allies: Stability and Structure in Arms Race Models." *Journal of Peace Science* 1 (1978): 45–54.

_____. "Response to Sudden Shifts in a Two-Nation Arms Race." *Behavioral Science* 22 (1977): 69–86.

Bechhoefer, Bernard G. *Postwar Negotiations for Arms Control.* Washington, D.C.: Brookings, 1961.

Bertram, Christoph. *The Future of Arms Control: Part II, Arms Control and Technological Change: Elements of a New Approach.* Adelphi Paper 146. London: International Institute for Strategic Studies, 1978.

Bloomfield, Lincoln P. "American Approaches to Military Strategy, Arms Control, and Disarmament: A Critique of the Postwar Experience." *Policy Studies Journal* 8 (1979): 114–19.

_____ and Cleveland Harlan. "A Strategy for the United States." *International Security* 2, no. 4 (1978): 32–55.

Bobrow, Davis B. "Arms Control through Communication and Information Regimes." *Policy Studies Journal* 8 (1979): 60–65.

Brame, Steven J., Morton D. Davis, and Philip D. Straffin. "The Geometry of the Arms Race." *International Studies Quarterly* 23, no. 4 (1979): 567–88.

Burns, Richard D. *Arms Control and Disarmament: A Bibliography.* Santa Barbara, Calif.: ABC-Clio, 1977.

Burt, Richard. "Reassessing the Strategic Balance." *International Security* 5 (1980): 37–52.

_____, ed. *Arms Control and Defense Postures in the 1980s.* Boulder, Colo.: Westview, 1982.

Cane, John W. "The Technology of Modern Weapons for Limited Military Use." *Orbis* 22, no. 1 (1978): 217–26.

Carnesale, Albert, et al. *Living with Nuclear Weapons.* New York: Bantam, 1983.

Clark, Michael and Marjorie Mowlam. *Debate on Disarmament.* Boston: Routledge & Paul, 1982.

Clark, Ronald W. *The Greatest Power on Earth: The International Race for Nuclear Supremacy.* New York: Harper, 1981.

Clarke, Duncan L. *The Politics of Arms Control: The Role and Effectiveness of the U.S. Arms Control and Disarmament Agency.* Glencoe, Ill.: Free Pr., 1979.

Collins, John M. *U.S.-Soviet Military Balance: Concepts and Capabilities 1960-1980.* New York: McGraw-Hill, 1980.

Dougherty, James E. *How to Think about Arms Control and Disarmament.* New York: Crane, Russak, 1973.

Dumas, Lloyd J., ed. *The Political Economy of Arms Reduction.* Boulder, Colo.: Westview, 1982.

Dupuy, Trevor N. and Gay M. Hammerman. *A Documentary History of Arms Control and Disarmament.* New York: T. N. Dupuy Associates, 1973.

Enthoven, Alain C. and K. Wayne Smith. *How Much Is Enough? Shaping the Defense Program, 1961-1969.* New York: Harper, 1971.

Feld, Bernard T., ed. *Impact of New Technologies on the Arms Race: A Pugwash Monograph.* Cambridge, Mass.: MIT Pr., 1971.

Gallois, Pierre. *The Balance of Terror: Strategy for the Nuclear Age.* Boston: Houghton, 1961.

Gelb, Leslie. "The Future of Arms Control: A Glass Half Full." *Foreign Policy* 36 (1979): 21-32.

Gillespie, John V., Diana A. Zinnes, and Michael R.Rubison."Accumulation in Arms Race Models: A Geometric Lag Perspective." *Comparative Political Studies* 10, no. 4 (1978): 475-96.

_____, et al., "Deterrence and Arms Races: An Optimal Control Systems Model." *Behavioral Science* 4 (1979): 250-62.

Goldblat, Jozef. *Agreements for Arms Control: A Critical Survey.* London: Taylor & Francis, 1982.

Gray, Colin S. *The Soviet-American Arms Race.* Lexington, Mass.: Lexington Books, 1976.

Hamblin, Robert L., et al., "Arms Races: A Test of Two Models." *American Sociological Review* 42, no. 2 (1977): 338-54.

Hill, Kim Quaile. "Domestic Politics, International Linkages, and Military Expenditures." *Studies in Comparative International Development* 13, no. 1 (1978): 38-59.

Hill, Walter W. "A Time-Lagged Richardson Arms Race Model." *Journal of Peace Science* 3, no. 1 (1978): 55-62.

Holloway, David. *The Soviet Union and the Arms Race.* New Haven, Conn.: Yale Univ. Pr., 1983.

Holst, Johan J. and Uwe Nerlich, eds. *Beyond Nuclear Deterrence: New Aims, New Arms.* New York: Crane, Russak, 1977.

Hunter, John E. "Mathematical Models of a Three-Nation Arms Race." *Journal of Conflict Resolution* 24, no. 2 (1980): 241-52.

Intriligator, Michael D. and D. C. Brito, "Formal Models of Arms Races." *Journal of Peace Science* 2, no. 1 (1976): 77-88.

Kelleher, Catherine M. "The Present as Prologue: European and Theater Nuclear Modernization." *International Security* 5, no. 4 (1981): 150-68.

Kemp, Anita. "A Path Analytic Model of International Violence." *International Interactions* 4, no. 1 (1978): 53–85.

Kupperman, Robert H. and Harvey A. Smith. "Formal Models of Arms Races: Discussion." *Journal of Peace Science* 2, no. 1 (1976): 89–96.

Legault, Albert and George Lindsey, eds. *The Dynamics of the Nuclear Balance*. Ithaca, N.Y.: Cornell Univ. Pr., 1974.

Lucier, Charles E. "Changes in the Values of Arms Race Parameters." *Journal of Conflict Resolution* 23, no. 1 (1979): 17–40.

Luck, Edward C. "The Arms Trade." *Proceedings of the Academy of Political Science* 32, no. 4 (1977): 170–83.

Luttwak, Edward. *The U.S.-U.S.S.R. Nuclear Weapons Balance*. Beverly Hills, Calif.: Sage, 1974.

Majeski, Stephen J. and David L. Jones. "Arms Race Modeling: Causality Analysis and Model Specification." *Journal of Conflict Resolution* 25, no. 2 (1981): 259–88.

Moll, Kendall D. and Gregory M. Moll. "Arms Race and Military Expenditures Models: A Review." *Conflict Resolution* 24, no. 1 (1980): 153–85.

Morgan, Patrick M. "Arms Control: A Theoretical Perspective." *Policy Studies Journal* 8 (1979): 105–14.

Moulton, Harland B. *From Superiority to Parity: The United States and the Strategic Arms Race, 1961–1971*. Westport, Conn.: Greenwood, 1973.

Myrdal, Alva. *The Game of Disarmament: How the United States and Russia Run the Arms Race*. New York: Pantheon, 1977.

Nincic, Miroslav. *The Arms Race: The Political Economy of Military Growth*. New York: Praeger, 1982.

Ostrom, Charles W., Jr. "A Reactive Linkage Model of the U.S. Defense Expenditure Policymaking Process." *American Political Science Review* 72, no. 3 (1978): 941–57.

Perry, William J. "Advanced Technology and Arms Control." *Orbis* 26, no. 2 (1982): 351–59.

Platt, Alan and Lawrence Weiler, eds. *Congress and Arms Control*. Boulder, Colo.: Westview, 1978.

Ranger, Robert. *Arms and Politics 1958–1978*. New York: Macmillan, 1979.

Rathjens, G. W. "Changing Perspectives in Arms Control." *Daedalus* 104, no. 3 (1975): 201–14.

Rattinger, Hans. "Armaments, Detente, and Bureaucracy: The Case of the Arms Race in Europe." *Journal of Conflict Resolution* 19, no. 4 (1975): 571–95.

————. "From War to War to War: Arms Races in the Middle East." *International Studies Quarterly* 20, no. 2 (1976): 501–31.

Richelson, Jeffrey T. "Evaluating the Strategic Balance." *American Journal of Political Science* 24, no. 4 (1980): 779–803.

Rosen, Steven, ed. *Testing the Theory of the Military Industrial Complex*. New York: Lexington Books, 1973.

Russett, Bruce M. and Bruce Blair, eds. *Progress in Arms Control?* San Francisco: Freeman, 1979.

Schelling, Thomas. "A Framework for the Evaluation of Arms Control Proposals." *Daedalus* 104, no. 5 (1975): 175–85.

Schrodt, Philip A., John V. Gillespie, and Diana A. Zinnes. "Parameter Estimation by Numerical Minimization Methods." *International Interactions* 4, no. 4 (1978): 279–302.

Smith, Theresa C. "Arms Race, Instability and War." *Journal of Conflict Resolution* 24, no. 2 (1985): 253–84.

Sorenson, David S. "Modeling the Nuclear Arms Race: A Search for Bounded Stability." *Journal of Peace Science* 4, no. 2 (1980): 169–85.

"The Soviet-American Arms Race and Arms Control." *Current History: A World Affairs Journal* 82 (1983).

Stockholm International Peace Research Institute. *The Arms Race and Arms Control*. London: Taylor & Francis, 1982.

_____. *Tactical Nuclear Weapons: European Perspectives*. London: Taylor & Francis, 1978.

Tsipis, Kosta. *Arsenal: Understanding Weapons in the Nuclear Age*. New York: Simon & Schuster, 1983.

U.S. Arms Control and Disarmament Agency. *Arms Control and Disarmament Agreements: Texts and History of Negotiations*. Washington, D.C.: Govt. Print. Off., 1980.

_____. *Documents and Disarmament*, 1961– . Washington, D.C.: Govt. Print. Off., 1961– .

Wallace, Michael D. "Arms Races and Escalation: Some New Evidence." *Journal of Conflict Resolution* 23, no. 1 (1979): 3–16.

Zinnes, Diana A., et al. "An Optimal Control Model of Arms Races." *American Political Science Review* 71, no. 1 (1977): 226–44.

Socioeconomic and Political Consequences of the Arms Race

Adams, Gordon. *The Politics of Defense Contracting: The Iron Triangle*. Brunswick, N.J.: Transaction Books, 1982.

Delineating the relationships among the Executive Branch (Department of Defense, NASA, Department of Energy), congressional appropriations committees (and members of Congress from defense-dependent districts), and private firms, institutes, and unions involved in defense production, this book examines the domestic arms race. Arising from these relationships are pressures on policy and weapon acquisition that result in increased (unnecessary) military expenditures and weapon systems of questionable utility and effectiveness. Case studies on eight major contractors are provided that emphasize their influence on military policy. This work is generally critical of the military-congressional-industrial complex and recommends various steps to democratize the triangle of interests and actions that forms the basis of defense in the United States.

Dellums, Ronald V., R. H. Miller and H. Lee Halterman. Edited by Patrick O'Heffernan. *Defense Sense: The Search for a Rational Military Policy*. Cambridge, Mass.: Ballinger, 1983.

This reader presents an abridged version of papers and testimony at a Special Congressional Ad Hoc Hearing on the Full Implications of the Military Budget in 1982. Six questions are addressed: (1) Weapons for what purposes?; (2) Foreign policy, national security, and the military budget: for what ends?; (3) Reagan's military budget: where are we going?; (4) Military spending and the domestic economy: what's left to defend?; (5) Moral implications of the military budget: a citizen responsibility?; and (6) Where does this lead us? Presenters include politicians, former policymakers, strategists, military critics, and religious leaders. The intent of the book is to provide a wide-ranging critique of existing military and foreign policy and to advance alternate policy directions. This is a useful source for persons seeking a critical introduction to the debate over militarization and the arms race.

Dumas, Lloyd J. "Thirty Years of the Arms Race: The Deterioration of Economic Strength and Military Security." *Peace and Change* 4, no. 2 (1977): 3-9.

This article is representative of those critiques of the permanent war economy that see the arms race as leading to lessened national security and economic decay. It is argued that the size and composition of military spending has cumulative negative pressures on inflation, employment, trade, deficits, and spending priorities. Reduced spending, lower dependence on the arms race, and redirected resources will result in greater security and economic revitalization.

Eide, Asbjorn and Marek Thee, eds. *Problems of Contemporary Militarism.* New York: St. Martin's, 1980.

This book includes nineteen readings on the subject of militarism and its impact on attitudes, social institutions, and international relations. The readings are critical of militarism and militarization as global trends. Particular emphasis is given to a conceptual examination of the global structure of militarism, the impact of the military in the Third World, and the political and cultural threats associated with the spread of militarism. The general orientation is that current military spending and the arms race divert economic resources (thereby retarding social development), threaten domestic and international political stability, and represent counterproductive mechanisms in the search for global security. The strength of the book is its international perspective, particularly its analysis of the relationship between the policies and programs of the developed world and political instability among developing nations.

Kidron, Michael and Dan Smith. *The War Atlas. Armed Conflict—Armed Peace.* New York: Simon & Schuster, 1983.

This book consists of forty maps and cartograms that facilitate both description and analysis of various dimensions of the arms race. The first section includes data on the extent of war (casualties, geographical location), the export of civil wars, preparations or practicing for warfighting (such as military exercises and maneuvers), and global nuclear targets. The second section focuses on the size and composition of the U.S. and Soviet arsenals (nuclear, biological, and chemical) and measures of conventional military forces. The third section illustrates the internationalization of the arms race: foreign military installations, military advisors, deployment of American and Soviet armed forces, and so on. The following two sections include data on the economic dimensions of the arms race: proportional military spending, military related employment, arms trade and transfer, and delineation of major exporters and importers of weapons and military technology. The sixth section describes collateral damage; that is unintended consequences of the arms race. Included are maps on military governments, internal political repression and the military, nuclear accidents, and environmental effects of conventional war. The last section includes data on the readiness and reliability of existing military organization, the emergence of antiarms protests, and tabular data on military characteristics for 147 countries. This book is quite useful in its synthesis of data, creativity of presentation, and illustration of the varying dimensions of the arms race.

Melman, Seymour. *Pentagon Capitalism: The Political Economy of War*. New York: McGraw-Hill, 1970.

This work remains one of the most important critiques of the military in America. It examines the concentration and centralization of government military power. Melman argues that the emerging management of the Pentagon has become a power within itself resulting in threats to democratic decision making, reduced public spending for social needs, and long-term economic decay. Particular emphasis is given to the effects of management centralization (for example, as arising from changes initiated by Secretary McNamara) on the American industrial base, labor, research and development and export of armaments, the role of universities in military research, and the emergence of militarism within both private and public sectors. This book represents one of the earliest and strongest attacks on the ''folklore'' of defense spending and its purported economic efficiency, spin-off effects on innovation, employment benefits, and so on.

_____, ed. *The War Economy of the United States. Readings on Military Industry and Economy*. New York: St. Martin's, 1971.

This book represents an extension of the argument articulated in *Pentagon Capitalism*. Its primary focus is the socioeconomic consequences of the military-industrial complex. The topics covered remain important issues today: convertability of military industry to civilian (peaceful) purposes, the impact of the military on civilian political institutions, the diversion of economic resources and resulting misdirected spending priorities, and so on. The book concludes with a discussion of the means and social consequences of disarmament (reduced military spending and reordered political priorities).

United Nations. Department of Political and Security Council Affairs. *Economic and Social Consequences of the Arms Race and of Military Expenditures*. New York: United Nations, 1972 and 1977.

These two reports illustrate the Third World critique of the arms race. Essentially, it is argued that the arms race creates and reinforces a hierarchical military and political system that threatens global development. Conventional arms trade and transfer and the spiraling nuclear rivalry between the United States and Soviet Union are seen as draining economic resources, reducing productivity, facilitating militarism, and greatly increasing the costs of war. This critique leads to an emphasis on disarmament (reduced military spending with redirected use of resources) in both developed and developing countries. The updated 1977 report examines the problems and prospects of a continuing arms race versus disarmament along economic, military, social, and political dimensions. These reports should be viewed more as a reflection of the perspective of nonnuclear developing nations that emphasize nonviolent mechanisms of conflict resolution and global economic development than as a reference source for empirical data.

_____. UNESCO. Reports and Papers in the Social Sciences, Number 39. *Review of Research Trends and an Annotated Bibliography: Social and Economic Consequences of the Arms Race and Disarmament*. Paris: UNESCO, 1978.

This report is a useful reference guide for the person interested in the literature on domestic international implications of the arms race. A brief introduction and discussion of issues is followed by an annotated bibliography of almost 200 studies and papers on the arms race. The review discussion focuses on the general perspectives and findings on the consequences of the arms race and disarmament, arms transfers, and research in socialist countries. The bibliographic material is structured around six topics: (1) the military establishment, military industry and society; (2) arms race, disarmament and the economy; (3) arms trade and military assistance; (4) military regimes in the third world and their impact on society; (5) military research and developments and impact on scientific institutions; and (6) consequences of disarmament and of the arms race for the international system and its processes.

Abrahamson, Bengt. *Military Professionalism and Political Power*. Beverly Hills, Calif.: Sage, 1972.

Albrecht, Ulrich, D. Lock and H. Wulf. "Armaments and Underdevelopment." *Bulletin of Peace Proposals* 5 (1974): 173–85.

_____. "Militarization, Arms Transfer and Arms Production in Peripheral Countries." *Journal of Peace Research* 12 (1975): 195–212.

Barnet, Richard. "U.S.–Soviet Relations: The Need for a Comprehensive Approach." *Foreign Affairs* 57 (1979): 779–95.

Benoit, Emile. *Defense and Economic Growth in Developing Countries*. Lexington, Mass.: Lexington Books, 1973.

_____ and Kenneth Boulding. *Disarmament and the Economy*. New York: Harper, 1963.

Blechman, Barry M. and Stephen S.Kaplan, eds.*The Use of the Armed Forces as a Political Instrument*. Washington, D.C.: Brookings, 1976.

_____, Jane E. Nolan, and Alan Platt. "Pushing Arms." *Foreign Policy* 46 (1982): 138–54.

Bolton, Roger E., ed. *Defense and Disarmament: The Economics of Transition*. Englewood Cliffs, N.J.: Prentice-Hall, 1966.

Boskey, Bennet and Mason Willrich, eds. *Nuclear Proliferation: Prospects for Control*. New York: Dunellen, 1970.

Boston Study Group. *The Price of Defense: A New Strategy for Military Spending*. New York: Times Books, 1979.

Brenner, Michael J. *Nuclear Power and Non-Proliferation: The Re-Making of U.S. Policy*. New York: Cambridge Univ. Pr., 1981.

"Burdens of Militarization." *International Social Science Journal* 25, no. 1 (1983).

Burt, Richard R. *Developments in Arms Transfers: Implications for Supplier Control and Recipient Autonomy*. Santa Monica, Calif.: Rand, 1977.

Cannizzo, Cindy, ed. *The Gun Merchants: Politics and Policies of the Major Arms Suppliers*. New York: Pergamon, 1980.

Chichilnisky, G. *The Role of Armament Flows in the International Market and in Development Strategies in a North-South Context*. New York: Columbia Univ. Pr., 1980.

Cioffi-Revilla, Claudio A. ''A COSP Catastrophe Model of Nuclear Proliferation.'' *International Interactions* 4 (1978): 191–224.

Cordesman, Anthony. ''U.S. and Soviet Competition in Arms Exports and Military Assistance.'' *Armed Forces Journal* (August 1981: 65–72.

DeMesquita, Bruce and William H. Riker. ''An Assessment of the Merits of Selective Nuclear Proliferation.'' *Journal of Conflict Resolution* 26 (1982): 283–306.

Doorn, Jacques van, ed. *Military Professions and Military Regimes*. Nouton: The Hague, 1969.

Drell, Sydney D. *Facing the Threat of Nuclear Weapons*. Seattle: Univ. of Washington Pr., 1983.

Dunn, Lewis A. and Herman Kahn. *Trends in Nuclear Proliferation, 1975–1995*. Croton-on-Hudson, N.Y.: Hudson Institute, 1976.

Durie, Sheila and Robert Edwards. *Fueling the Nuclear Arms Race: The Links between Nuclear Power and Nuclear Weapons*. London: Pluto Pr., 1982.

Eide, Asbjorn. ''Arms Transfer and Third World Militarization.'' *Bulletin of Peace Proposals* 8 (1977): 220–32.

————. ''The Transfer of Arms to Third World Countries and Their Internal Uses.'' *International Social Science Journal* 28 (1976): 307–25.

Epstein, William. *The Last Chance: Nuclear Proliferation and Arms Control*. New York: Free Pr., 1976.

————. ''Why States Go—and Don't Go—Nuclear.'' *Annals of the American Academy of Political and Social Science* 430 (1977): 16–28.

Falk, Richard and Samuel Kim, eds. *The War System: An Interdisciplinary Approach*. Boulder, Colo.: Westview, 1980.

Fallows, James. *National Defense*. New York: Random, 1981.

Farley, Philip J., Stephen S. Kaplan, and William H. Lewis. *Arms across the Sea*. Washington, D.C.: Brookings, 1978.

Feit, Edward. *The Armed Bureaucrats: Military-Administrative Regimes and Political Development*. Boston: Houghton, 1973.

Fidel, Kenneth, ed. *Militarism in Developing Countries*. Brunswick, N.J.: Transaction Books, 1975.

Fox, John Ronald. *Arming America: How the U.S. Buys Weapons*. Cambridge: Harvard Univ. Pr., 1974.

Frank, Lewis A. *The Arms Trade in International Relations*. New York: Praeger, 1979.

Galtung, Johan. ''A Structural Theory of Imperialism.'' *Journal of Peace Research* 8 (1971): 81–118.

Gansler, Jacques. *The Defense Industry*. Cambridge, Mass.: MIT Pr., 1980.

Goheen, Robert F. "Problems of Proliferation: U.S. Policy and the Third World." *World Politics* 35 (1983): 194–215.

Graham, Thomas W. and Ridgely C. Evers. *Bibliography: Nuclear Proliferation.* Washington, D.C.: Govt. Print. Off., 1978.

Greenwood, Ted, Harold Feiverson, and Theodore Taylor. *Nuclear Proliferation: Motivations, Capabilities and Strategies for Control.* New York: McGraw-Hill, 1977.

Grimmett, Richard F. *Trends in Conventional Arms Transfers to the Third World by Major Suppliers, 1976–1983.* Washington, D.C.: Congressional Research Service, 1984.

Gutteridge, William. *Armed Forces in New States.* London: Oxford Univ. Pr., 1962.

_____. *Military Institutions and Power in the New States.* London: Oxford Univ. Pr., 1964.

Hammond, Paul Y., David J. Louscher, and Michael D. Salomon. "Controlling U.S. Arms Transfers: The Emerging System." *Orbis* 23 (1979): 317–52.

Harkavy, Robert E. *The Arms Trade and International Systems.* Cambridge, Mass.: Ballinger, 1975.

Hitch, Charles J. and Roland N. McKean. *The Economics of Defense in the Nuclear Age.* Cambridge, Mass.: Harvard Univ. Pr., 1967.

Howe, Russell W. *Weapons: The International Game of Arms, Money, and Diplomacy.* Garden City, N.J.: Doubleday, 1980.

International Peace Research Association. "The Impact of Militarization on Development and Human Rights." *Bulletin of Peace Proposals* 9 (1978): 170–82.

Janowitz, Morris. *The Military in the Political Development of New Nations: An Essay in Comparative Analysis.* Chicago: Univ. of Chicago Pr., 1964.

Johnson, John J., ed. *The Role of the Military in Underdeveloped Countries.* Princeton, N.J.: Princeton Univ. Pr., 1962.

Jolly, Richard, ed. *Disarmament and World Development.* Oxford: Pergamon, 1978.

Jones, Rodney, ed. *Small Nuclear Forces and U.S. Security Policy: Threats and Potential Conflicts in the Middle East and South Asia.* Lexington, Mass.: Lexington Books, 1984.

_____ and Steven A. Hildreth. *Modern Weapons and Third World Powers.* Boulder, Colo.: Westview, 1984.

Kaldor, Mary. "The Military in Development." *World Development* 4 (1976): 459–82.

_____ and Asbjorn Eide, eds. *The World Military Order: The Impact of Military Technologies in the Third World.* London: Macmillan, 1979.

Katz, James E. and Onkar S. Marwah, eds. *Nuclear Power in Developing Countries.* Lexington, Mass.: Heath, 1982.

Kegley, Charles W., Jr. and Eugene R. Wittkopf, eds. *The Nuclear Reader: Strategy, Weapons, and War.* New York: St. Martin's, 1985.

_____. *World Politics: Trends and Transformation.* New York: St. Martin's, 1985.

Kemp, Geoffrey, Robert Pfatzgraaff, and Uri Ra'anan. *The Other Arms Race: New Technologies and Non-Nuclear Conflict*. Lexington, Mass.: Lexington Books, 1975.

Kennedy, Gavin. *The Military in the Third World*. London: Duckworth, 1974.

Klare, Michael T. "The Political Economy of Arms Sales." *The Bulletin of the Atomic Scientists* 9 (1976): 11–16.

_____ and Cynthia Arnson. *Supplying Repression*. 2nd ed. Washington, D.C.: Institute for Policy Studies, 1981.

Knorr, Klaus. "On the International Use of Military Force in the Contemporary World." *Orbis* 21, (1977): 5–27.

Landgren-Backstrom, Signe. "The Transfer of Military Technology to Third World Countries." *Bulletin of Peace Proposals* 8 (1977): 110–20.

Lapp, Ralph E. *The Weapons Culture*. New York: Norton, 1968.

Lawrence, Robert M. and Joel Larus, eds. *Nuclear Proliferation: Phase II*. Lawrence, Kan.: Univ. Pr. of Kansas, 1974.

Lefever, Ernest W. *Nuclear Arms in the Third World: U.S. Policy Dilemma*. Washington, D.C.: Brookings, 1979.

Lens, Sidney. *The Military-Industrial Complex*. London: Stanmore Pr., 1971.

Little, Roger W., ed. *Handbook of Military Institutions*. Beverly Hills, Calif.: Sage, 1971.

Lock, Peter. "Armament Dynamics: An Issue in Development Strategies." *Alternatives* 6 (1980): 157–78.

_____ and Herbert Wulf. "Consequences of Transfer of Military-Oriented Technology on the Development Process." *Bulletin of Peace Proposals* 8 (1977): 127–36.

Long, Franklin A. and Judith Ruppy. *The Genesis of New Weapons: Decision Making for Research and Development*. Oxford: Pergamon Press, 1980.

Lumsden, Malvern. "Global Military Systems and the New International Economic Order." *Bulletin of Peace Proposals* 9 (1978): 30–34.

Mallmann, Wolfgang. "Arms Transfers to the Third World: Trends and Changing Patterns in the 1970s." *Bulletin of Peace Proposals* 10 (1979): 301–7.

Meyer, Stephen M. *Nuclear Proliferation: Models of Behavior. Choice and Decision*. Chicago: Univ. of Chicago Pr., 1983.

Newman, Stephanie and Robert Harkavy. *Arms Transfers in the Modern World*. New York: Praeger, 1979.

Oeberg, Jan. "Arms Trade with the Third World as an Aspect of Imperialism." *Journal of Peace Research* 12 (1975): 213.

Office of Technology Assessment. *Nuclear Proliferation and Safeguards*. Washington, D.C.: National Technical Information Service, 1977.

Pierre, Andrew J. "Arms Sales: The New Diplomat." *Foreign Affairs* 60 (1981–82): 266–86.

Potter, William C. *Nuclear Power and Nonproliferation: An Interdisciplinary Perspective*. Cambridge, Mass.: Oelgeschlager, Gunn & Hain, 1982.

Quester, George. *The Politics of Nuclear Proliferation*. Baltimore: Johns Hopkins Univ. Pr., 1973.

_____, ed. *Nuclear Proliferation: Breaking the Chain*. Madison, Wisc.: Univ. of Wisconsin Pr., 1981.

_____. "The Statistical '*n*' of '*n*th' Nuclear Weapons States." *Journal of Conflict Resolution* 27 (1983): 161–79.

Rossiter, Caleb S. *U.S. Arms Transfers to the Third World: The Implications of Sophistication.* Washington, D.C.: Congressional Research Service, 1982.

Sampson, Anthony. *The Arms Bazaar: The Companies, the Dealers, the Bribes: From Vickers to Lockheed.* London: Hudder & Stoughton, 1977.

Schoettle, Enid. *Postures for Non-Proliferation: Arms Limitation and Security Policies to Minimize Nuclear Proliferation.* London: Taylor & Francis, 1969.

Senghaas, Dieter. "Military Dynamics in the Contemporary Context of Periphery Capitalism." *Bulletin of Peace Proposals* 8 (1977): 103–9.

Stanley, John and Maurice Pearton. *The International Trade in Arms.* London: Chatto & Windus, 1972.

Stevenson, Paul. "The Military-Industrial Complex: An Examination of the Nature of Corporate Capitalism in America." *Journal of Political and Military Sociology* 1 (1973): 247–59.

Stockholm International Peace Research Institute. *The Arms Trade with the Third World.* Stockholm: Almquist & Wiskell, 1971.

_____. *World Armament and Disarmament. SIPRI Yearbook 1977.* Stockholm: SIPRI, 1977.

Thayer, George. *The War Business: The International Trade in Armaments.* London: Weidenfeld & Nicholson, 1969.

Thee, Marek. "The Dynamics of the Arms Race, Military R&D, and Disarmament," *International Social Science Journal* 30 (1978): 904–25.

United States Arms Control and Disarmament Agency. *The International Transfer of Conventional Arms.* Report to the Congress, 93rd Cong., 2nd sess. Washington, D.C.: Govt. Print. Off., 1974.

U. S. Congress. Congressional Budget Office. *Budgetary Cost Savings to the Department of Defense Resulting from Foreign Military Sales.* Washington, D.C.: Govt. Print. Off., 1976.

_____. House. Committee on Foreign Affairs. Subcommittee on International Security and Scientific Affairs. *Changing Perspectives on U.S. Arms Transfer Policy.* 97th Cong., 1st sess. Washington, D.C.: Govt. Print. Off., 1981.

_____. Committee on International Relations. Subcommittee on Inter-American Affairs. *Arms Trade in the Western Hemisphere.* Hearings. 95th Cong., 2nd sess. Washington, D.C.: Govt. Print. Off., 1978.

_____. Senate. Committee on Foreign Relations. *Arms Sales and Foreign Policy.* 90th Cong., 1st sess. Washington, D.C.: Govt. Print. Off., 1967.

_____. *U.S. Conventional Arms Transfer Policy.* 96th Cong., 2nd sess. Washington, D.C.: Govt. Print. Off., 1980.

_____. Subcommittee on Foreign Assistance. *Foreign Assistance Authorization: Arms Sales Issues.* Hearings. 94th Cong., 1st sess. Washington, D.C.: Govt. Print. Off., 1975.

Varynynen, Raimo. *Arms Trade, Military Aid and Arms Production.* Basel: Herder Verlag, 1973.

_____. "Economic and Political Consequences of Arms Transfers to the Third World." *Alternatives* 6 (1980): 131–55.

Wolpin, Miles D. *Military Aid and Counter-Revolution in the Third World.* Lexington, Mass.: Lexington Books, 1972.
Yarmolinsky, Adam. *The Military Establishment: Its Impact on a Society.* New York: Harper, 1971.

Part 3

DETERRENCE AND THE FUTURE

CHAPTER 7

The Debate over Deterrence and Strategic Theory

I n 1946, Bernard Brodie wrote:

> Thus far the chief purpose of our military establishment has been to win wars. From now on its chief purpose must be to avert them.[1]

Much of the nuclear debate can be reduced to whether one accepts or rejects Brodie's thesis that the advent of nuclear weapons requires a shift from a war-fighting to a war-deterring orientation. Can nuclear weapons be used strategically and successfully in war, or would their use be so disastrous that the prospect should dissuade nations from escalating any conflict to the point where their threatened employment will occur?

During the 1970s and 1980s these questions have been much more under debate than before. Several factors have contributed to the intensification of the debate and its increasingly public nature. First, from the time of the Nixon administration to the present, nuclear policies have more conspicuously stressed counterforce options. Also, counterforce options themselves have become more technically feasible as a result of increased missile accuracy. Both the shifts in policies and the refinements in weapons, in turn, were influenced by changes in strategic theory. The 1970s and 1980s, in brief, reflect the shifts in the influence of the two main schools of strategy. This chapter provides historical background to this debate as it has occurred within and outside the strategic community.

In order to understand this fundamental issue in the nuclear debate, several distinctions need to be kept in mind. In brief, deterrence theory and strategic theory need to be distinguished from each other and from both policy declarations and operational scenarios. We discussed policy in Chapter 3; it concerns official statements about the current role of nuclear weapons in the pursuit of national security. We discussed scenarios in Chapter 4; they

---

1. Bernard Brodie, ed., *The Absolute Weapon: Atomic Power and World Order* (New York: Harcourt, 1946), p. 76.

are the actual plans for how to use currently available nuclear weapons, if required. Deterrence and strategy are theories that can inform the design of declaratory policy and the manner (scenario) of preferred use. Deterrence and strategy, moreover, are distinct theories. Strategic theory is the broader term; it refers to the perspectives on what a nation's overall approach to security should be (for example, it includes economic, diplomatic, and military components). Deterrence theory is one possible element in a strategic view; it refers to the capacity of perceived strengths to dissuade other nations from engaging in certain activities (it includes the threat of economic sanctions, diplomatic rebuff, and military response).

With regard to strategic theory and deterrence theory, one can reject either or both as pseudoscience, or one can defend or criticize any of the numerous versions of each. This chapter is structured around the ways in which these options have been played out, both within the strategic community and in relation to governmental policy formulation. The chapter consists of four sections. The first three sections focus on the relation between strategists and politicians. Generally, strategists are civilians who study security issues from a social-scientific perspective and formulate theories and models concerning preferred approaches. Politicians, in both the executive and legislative branch of government, set policy positions that may or may not be informed by strategic theory.

The first section of this chapter covers the period from 1945 to 1960, during which policymaking was largely independent of any significant input by the strategic community. The second section covers the 1960s when not only did the strategic community begin to shape governmental policy, but one particular strategic school monopolized governmental planning. The third section traces the debate during the 1970s and 1980s between two main schools within the strategic community and discusses how government planning shows partial reliance on both schools. Finally, the fourth section surveys the perspectives on deterrence and strategy that come from outside the strategic community. These writers often view the issue more from the perspectives of philosophy and religion than from political science and international relations.

One further observation needs to be made in distinguishing this chapter from the next and final one. Chapter 7 focuses on the defenses and criticisms of various deterrence and strategic theories. Chapter 8 will focus on alternative models. Some writers are considered in both chapters—their criticism is cited in this chapter and their alternatives in the next.

The Independence of Policymaking from the Strategic Community: 1945 to 1960

Although initial U.S. nuclear policy was simply an extension of strategic bombardment, it was shaped by conventional strategy rather than by a theory concerning nuclear weapons per se. Only after the destruction of Hiroshima and Nagasaki did any theorists really think through the strategic impact of nuclear weapons. Brodie (1946) expressed succinctly the facts with which strategic theory had to come to terms:

> Everything about the atomic bomb is overshadowed by the twin facts that it exists and its destructive power is fantastically great.[2]

Two schools quickly emerged and have continued their disagreement ever since. Given Brodie's facts, one saw nuclear weapons as a great asset in war fighting, and the other saw the use of nuclear weapons as a terrible bane.

During the Truman and Eisenhower administrations, the view that nuclear weapons would aid war fighting prevailed. In part, the Cold War between the United States and the Soviet Union supported continued military preparedness, and in this context atomic weapons were viewed as tried and proven additions to the U.S. arsenal. Moreover, the number and, initially, the yield of nuclear weapons were sufficiently low that these weapons were often viewed more as a highly efficient supplement than as the predominant and adequate group of weapons. LeMay's mission at SAC headquarters was to find a target for every atomic bomb now or soon to be available. Viewing the atomic bomb as useful but inadequate for ending a major conflict, Blackett and Bush expected the next war would be a long one.

In contrast, the specifically nuclear theorists wanted to avoid war because of the nature of atomic weapons. This line of reasoning can be found in Viner, the Chicago economist who wrote perhaps the first and clearly one of the most succinct and incisive essays on what would later be termed deterrence theory, in Brodie (1946), who was influenced by Viner and later taught Kaufmann (who, through his influence on McNamara, was to shape nuclear policy in the 1960s), and in Borden, who was influenced by Brodie.

While desiring to avert war, these theorists thought through the options for developing and using nuclear weapons for at least two reasons. First, they believed that such forces could function for deterrence. Second, they believed that despite the destructiveness of atomic weapons, there was little prospect for what is now termed arms control. The initial postwar efforts at nuclear arms control had failed partly because the Soviets wanted to be able to match the U.S. achievement, and partly because the United States wanted to maintain its clear nuclear dominance. Especially after Soviet entry into the nuclear club, governmental motivation for arms control was low. Sharing their pessimism over arms control yet rejecting their orientation toward deterrence, Gray, one of the

2. *Ibid.*, p. 52.

leading present-day advocates of a war-fighting orientation, accurately notes regarding the work of Viner, Brodie, and Borden that in their early theorizing is the initial discussion of

> deterrence theory and strategy, limited war, counterforce as opposed to countervalue strategies, secure as opposed to insecure retaliatory forces (the first-strike, second-strike distinction), the irrelevance of a capacity for wartime mobilization, the promise of ballistic missiles, and the political inutility of a quantitative advantage in a nuclear arms race.[3]

The theory of this initial strategic community had almost no policy impact and had to be rediscovered or reinvented by later theorists. In fact, Congress (1949 and 1951) only occasionally scrutinized the strategy behind early U.S. atomic forces and the executive branch relied on the National Security Council.

While U.S. nuclear policy was moving from containment to massive retaliation, strategic theory was largely at a standstill. Several factors for this lull can be noted. First, since government was not relying on strategists, that community did not grow. Also, universities had not yet developed sufficiently the kinds of programs that would produce increased numbers of civilian scholars in this area. Furthermore, because the basic positions had been laid out and a public forum for their continued debate was not available, there was neither an inclination nor an audience for the repetition of these views. From 1946 to 1956, Gray claims, "no book-length, wide-ranging works of stature were published on strategic theory.[4]

Despite the basic accuracy of Gray's observation, some truly seminal articles and studies did appear toward the close of this ten-year period, largely representing strategic thought spawned in reaction to the policy of massive retaliation. Especially relevant in this regard are Brodie (1954) and Wohlstetter (1954). The truly influential works, however, were written in the late 1950s and early 1960s and, despite their many differences, formed the dominant perspective within the strategic community and succeeded in shaping U.S. nuclear policy during the 1960s.

The Influence of the War-Deterring Strategic School on Policymaking: 1960s

While virtually all theorists who see nuclear weapons as an effective element in achieving national security agree that nuclear weapons help deter war, they disagree over the types and numbers needed and over their actual use. Much of this debate pertains to the theory of limited war

---

3. Colin S. Gray, *Strategic Studies and Public Policy: The American Experience* (Lexington, Ky.: Univ. Pr. of Kentucky, 1982), p. 29.

4. *Ibid.*, p. 49.

and the theory of arms control. The theory of limited war was largely hammered out during the late 1950s and the theory of arms control emerged in the late 1960s. Within the dominant school that emerged, it was assumed, against the policies of the prior two administrations, that deterrence was most likely under conditions of a balance of terror. In part, as the Soviet arsenal grew, it became increasingly infeasible to assume either side could use nuclear weapons without also incurring heavy losses. In a "no win" situation, many theorists reasoned that establishing equal forces on each side was more stable, since unequal forces might countenance the foolhardy belief that sheer quantitative superiority was sufficient for attaining victory. The quantitative shift from U.S. nuclear monopoly and then superiority to U.S.-Soviet nuclear parity parallels not just Soviet military buildup, but also a U.S. policy shift informed by the war-deterring school of the strategic community.

Freedman catches the irony of this liberal school when he notes how the very liberals who opposed conventional countervalue strikes came to accept nuclear countervalue strikes. He observes:

> Liberals had opposed strategic bombardment as a particularly uncivilized form of warfare. Within government they had resisted the continued effort to perfect thermonuclear means of destroying cities, attempting in the late 1940s and early 1950s to divert research on nuclear weapons to designs for tactical, battlefield use. Outside government they had deplored Western reliance on such weapons and had spoken with gloom and despondency of the future of mankind in the absence of effective international measures to promote disarmament. To turn the capacity for city destruction into a virtue and to use this as a foundation for peace and stability was quite perverse. Yet by the mid-1960s this approach had become almost the party "line" for liberals, with few dissenters.[5]

The virtue of this position is that it avoided the intent or capability of a counterforce first strike: deterrence would be achieved by each side having invulnerable retaliatory forces for a countervalue second strike should the other side launch a first strike. This deterrence theory grew out of a limited war theory and led to arms control theory. For writings in support of mutual assured destruction, see Brennan, Holton, Kaufman, King, and Schelling; criticisms of MAD are found in Brodie (1966), Kahn, and Nitze (1956).

Limited War Theory

Ever since World War II, Liddell Hart had been arguing for limited war. The horror of modern war—even before the atomic bomb—had become so great that means should be sought to limit and control it. Liddell

---

5. Lawrence Freedman, *The Evolution of Nuclear Strategy* (New York: St. Martin's Pr., 1981), p. 195.

Hart's contribution was to convert this sentiment into a sound strategic theory. As far as the influence of limited war theory on nuclear strategy is concerned, the development of the hydrogen bomb with its awesome destructiveness was even more decisive for many in rejecting total war as an option.

Significantly, Brodie later admitted to Liddell Hart that he had become one of his followers early in 1952 right after the United States tested the first thermonuclear weapon. This view can be found in public essays by Brodie (1954) that followed shortly afterwards and is quite conspicuous in the hallmark text of war-deterring strategic theory, *Military Policy and National Security* edited by Kaufmann (1956). Osgood and Kissinger later resounded this argument in their own influential contributions to limited war theory. Typically, this school favored use of tactical nuclear weapons in local wars. In opposition, King wanted to separate, not join, limited wars and nuclear weapons, and Buzzard favored limited war by graduated deterrence.

Despite the dominance of the war-deterring school, the war-fighting school had its representatives at this time, too. Gray, though noted for his recent call for a theory of victory, had important and likeminded predecessors, specifically Nitze who argued that there were important senses in which one could win a nuclear war.

Readers wanting the policy perspective on this issue as well as annotations on several of the key texts on limited war should consult Chapter 3.

Arms Control Theory

In the 1940s and 1950s the efforts to eliminate or regulate nuclear weapons were more a reaction to their destructiveness than an element in strategic theory. As the Soviet arsenal grew, some strategists rethought the relation between nuclear weapons and deterrence. The arms control theory that resulted rests on two key premises. First, it assumes a major nuclear war would be catastrophic. Second, it assumes U.S. and Soviet arsenals can be held to levels at which a stalemate exists. In other words, arms control rejects efforts at both the type of superiority needed for an aim of victory in nuclear war and the type of inferiority resulting from a program of unilateral nuclear disarmament.

Arms control became central to the war-deterring schools of the strategic community with the 1960 special issue of *Daedalus* on arms control (see Holton). The following year, three key texts appeared: Brennan, Bull, and Halperin. Writings on the origins of arms control theory go back to Amster (1954, 1956) and Sherwin. To examine the strategy of stable conflict, with which the pursuit of arms control and parity are closely linked, see Schelling (1960 and 1961). To get a sense of the amount of literature on arms control, see Burns (1977).

Arms control theory has had success relative to avoiding both inferiority and superiority. With respect to inferiority (the call to reduce nuclear stockpiles), the relatively stable relation achieved between the United States and the Soviet Union silenced many who had advocated disarmament as the way to avoid nuclear war.[6] With respect to superiority (the call to increase nuclear stockpiles), negotiation of and compliance with arms control treaties made the advocates of increasing armaments appear to desire war-fighting over war-deterring forces. The success of advocates of arms control theory is reflected in the fact that over sixteen international agreements to control or limit nuclear arms have been negotiated since 1959. The centerpiece of these efforts at arms control was SALT, the Strategic Arms Limitation Talks. These U.S.-Soviet negotiations began in November 1969, and SALT I was completed in May 1972 with the ABM Treaty and the interim agreement. The former is of unlimited duration and restricts each side on deployment of antiballistic missile systems. The latter lasted five years and was followed by SALT II. Both the interim agreement and SALT II, which was completed in 1979, placed limits on ICBM and SLBM launches. SALT II also placed limits on intercontinental-range bombers and established various sublimits, such as on multiple warhead launchers. Although SALT II has not received Senate ratification, both the United States and the Soviet Union abided by it until late 1986 when the United States openly exceeded the limits, claiming that the Soviet Union had been in violation in several areas.

Several sources provide good overviews on these and other aspects of arms control. See U.S. Arms Control and Disarmament Agency for texts of and background on U.S.-Soviet agreements, Stockholm International Peace Research Institute (1978) for a broader look (including multilateral agreements on all types of weapons), and United Nations (1970, 1976) for another broad and annually updated look. More specific sources on the debates during this period can be found in Yanarella (1977), Chayes and Wiesner, and Smith.

As a result of shifting tides in the strategic debate, the SALT process was succeeded by Strategic Arms Reduction Talks (START) in June 1982. In addition to congressional hearings, SALT had generated considerable public debate throughout the 1970s, beginning the current period in which proper deterrence and strategic theory have come to be debated publicly. The principal points in the debate concern the desirability of parity or balance rather than either inferiority or superiority. The attack on parity and arms control, however, has not primarily involved advocates of disarmament; they now generally support arms control as the best hope until, if ever, conditions are politically ripe for

6. *Ibid.*, pp. 199–207.

genuine reductions. The group not pleased by the success of arms control is the community of "defense-minded" or "revisionist" strategists—the war-fighting school, which began to achieve some influence on policy during the 1970s and is exerting even more influence in the 1980s.

The Debate between and Policy Influence of the War-Deterring and War-Fighting Strategic Schools: 1970s and 1980s

By the 1970s the strategic community was neither small nor unified. In terms of size, several factors had played important roles in increasing the number of people involved. Strategic studies became institutionalized. On the one hand, prestigious universities had developed programs in these areas and were influencing curricula across the country as a result of efforts by students from these programs to secure academic employment. On the other hand, because SALT was a large-scale and ongoing process, successive groups gained credentials as strategists. The military itself and civilian centers dependent on government contracts, such as Rand, also provided much training in strategic studies. As a result, the number of persons with credentials as strategic experts far exceeds the number of governmental positions for persons with such skills. In terms of unity, additional factors had worked to undercut it. Among these factors, Freedman cites in particular

> the perplexing growth in Soviet missile forces, the critique of arms race theories relying on overexuberant weapons designers and intelligence estimators, the uncertainties over Soviet intentions, the moral qualms, and the confusion of technological progress.[7]

Congress, the media, churches, and the popular press also became increasingly involved. No new consensus has emerged even within the strategic community. Nevertheless, while there are many subgroups it is convenient and accurate to divide the current strategic community into the war-deterring and war-fighting schools. Writers in the 1970s and 1980s usually endorse one of these two views whether they are strategic theorists or not, often taking a specific position on one of the current counterforce weapon or policy debates. Since much of this literature can be accessed by topic and date using the bibliographies of the various chapters of this book, the issues and works discussed here are rather selective. The unannotated bibliography for this section cites a fair number of the key books and articles of the 1970s and 1980s that argue for and from one of the currently competing strategic theories.

7. *Ibid.*, pp. 354–55.

While a great many topics have been debated within the strategic community during the 1970s and 1980s, two issues have attracted primary concern—counterforce weapons, especially when coupled with first-strike strategies, and NATO, particularly in terms of the relative roles for conventional and nuclear weapons. The debate on space weapons (Reagan's Strategic Defense Initiative), which is covered in Chapter 8 dealing with alternatives, emerges from the war-fighting school's argument that the counterforce influence achieved in the 1970s is still insufficient. As evidence this school cites the fact that it would be difficult for the United States to "win" a nuclear war. While it concedes that the United States is moving toward developing and deploying the requisite offensive weapons, it argues that the viability of nuclear war will also require defense of the homeland, which the United States has largely abandoned both in its active form in ABM systems and in the passive form of civil defense. This school's proposal for space weapons, while appearing as defensive, is also integral to damage limitation in case of war.

Counterforce Strategy

Although counterforce options have always been included in U.S. nuclear policy (see Chapter 3), the counterforce debate took a more strategic turn once parity was approximated. Some, like Van Cleave and Burnett, argued that changes were needed to make feasible the use of nuclear weapons in limited war fighting. During the second Nixon administration, this concern within the war-fighting school of the strategic community achieved policy influence.[8]

The personal roles of Henry Kissinger and, later, James Schlesinger were motivating forces for giving more influence to these views. Kissinger was one of the architects of limited war theory and saw possibilities for nuclear weapons to aid pursuit of national interest. Nuclear weapons, in his view, not only help prevent attack but also help wield influence through their mere existence, their threatened use, and perhaps even their actual use (especially tactical nuclear weapons). When James Schlesinger became Nixon's secretary of defense, he was the first civilian strategist to hold this office. Specifically, his work at Rand and elsewhere represented practical efforts in system analysis that were basically in line with Kissinger's thought on strategic theory. Kissinger, however, was not an orthodox exponent of the war-fighting school. He was not opposed to arms control. Throughout Nixon's first term he worked for SALT I. Still, after Nixon's reelection, the door was open for pursuing counterforce options to a greater degree than before. In January 1974 Schlesinger announced what has come to be known as "the Schlesinger doctrine." In line with war-fighting strategies, he called for more nuclear options.

8. *Ibid.*, p. 377.

Some criticisms, of course, followed. On the one hand, Congress (March and September 1974) became concerned. Some of this debate is also discussed in Chapter 5, especially in the Senate's *Analyses of the Effects of Limited Nuclear Warfare* (Senate, 1975) and *Effects of Limited Nuclear Warfare* (Senate, 1976). On the other hand, a few representatives of the war-deterring school expressed objections. Among these strategic critiques, see especially Greenwood and Nacht, Scoville (1974), Carter, and Pranger and Labrie.

While many of these writers repeated (somewhat ineffectively) old arguments, Brodie (1978) broke new ground in what was to be his last contribution to nuclear strategy. Besides his succinct and insightful review of the history of nuclear strategy, Brodie criticizes what he terms the Schlesinger-Lambeth school, and, in contradistinction to their call for more nuclear options, he observes:

> The notion that it is incontestably good to expand the chief executive's options is rather peculiar. For one thing, it runs directly counter to the basic tenets of constitutional government. . . . It is an old story that one way of keeping people out of trouble is to deny them the means for getting into it. We have put in the President's hand a huge military power because we believe that the country's security demands that we do so, and we are obliged to trust that he will use it wisely. But to expand that power simply for the sake of expanding his options is to push hard that obligatory trust.[9]

Brodie's final words repeated the message and position of restraint he had been advocating since 1946 and which since the 1960s had helped shape U.S. nuclear policy.

In contrast to the war-deterring school's criticisms, the war-fighting school initially minimized the significance of the shift. Gray writes:

> There is no major distinction between the Schlesinger doctrine and the preferred deterrence-thinking of most civilian strategists (as opposed to the subcommunity of arms controls) over the previous two decades.[10]

While the shift may not be great, Gray's characterization of the strategic community is inexact. While all along limited war theory had been a part of the civilian strategists' approach, so had arms control at least since 1960.

This debate spills over into the controversy about NATO (treated in the next section), but it was resurrected and intensified during the Carter administration with the approval of Presidential Directive 59. This directive largely picked up where the Scheslinger doctrine left off (see section on SIOP in Chapter 4, as well as relevant portions of Chapter 3).

---

9. Steven E. Miller, ed., *Strategy and Nuclear Deterrence: An "International Security" Reader* (Princeton, N.J.: Princeton Univ. Pr., 1984), pp. 18–19.

10. Gray, *Strategic Studies and Public Policy*, p. 147.

From a strategic perspective, the issue concerned whether a theory of victory in nuclear war was plausible.

Gray (1979, 1980) is most closely associated with the segment of the war-fighting school that wants a theory of victory. Gray argues that when the president's options reduce to total war (annihilation) or no war (surrender), nuclear policy is immoral. Using criteria for just war, he argues for the development of capabilities and policies for war-fighting at all levels and with the prospect for victory, meaning at the least sufficient elimination of the Soviet leadership and economy to preclude a postwar communist power threat and sufficient protection of U.S. leadership, population, and economy to insure U.S. postwar dominance.

Beres (1980, 1983) has been one of Gray's foremost critics. Beres argues against the likelihood of a Soviet counterforce first strike (and the United States need to defend against it), against the view that the Soviets are likely to be deterred more by the position of the war-fighting than by the war-deterring school (supporting a return to minimum deterrence), against the view that significant postwar survival can be insured, and against the war-fighting school's defiance of arms control.

The disagreement between Gray and Beres can be seen in their reactions over Presidential Directive 59. In back-to-back articles in *Parameters*, Gray (1981) supported and Beres (1981) criticized PD-59.

Whereas the arms control theory of the war-deterring school pursues security (meaning stability) through a balance of U.S. and Soviet forces, the theory of victory of the war-fighting school pursues security (meaning an eventual Soviet roll-back) through a superiority of U.S. over Soviet forces. This shift from symmetry of force requires two additional premises. First, superiority should lie on the side of the United States. U.S., but not Soviet, military leaders should be able to offer political leaders "a plausible theory of military victory for the solution of some otherwise intractable political problem."[11] Second, crisis management and negotiation work primarily under conditions in which one side is clearly stronger; in such situations the weak make concessions to the strong. Since the war-fighting school wants to be able to achieve political or if necessary military influence over other nations' behavior, it does not want nuclear forces to stand in such a relation that the United States is subject to self-deterrence because of the effects of use. The war-deterring school accepts self-deterrence of U.S. nuclear forces as the price of correlative Soviet self-deterrence of nuclear forces.

NATO Strategy

Since 1967 NATO has operated with a strategy of flexible response. This strategy, implemented by McNamara, was designed to have more

---

11. *Ibid.*, p. 153.

credibility than massive retaliation. Massive retaliation was viewed in the United States as not being believable. Since many in the United States felt the Soviets did not believe the United States would risk possible Soviet retaliation to a massive U.S. strike against the Soviet Union, some sought limited U.S. nuclear options in Europe that the Soviets would believe would be executed in some situations.

Flexible response increases U.S. nuclear options. The strategy of flexible response all along has involved both conventional and nuclear forces. Moreover, while it aims for deterrence, it also provides for "forward defense" (war-fighting). It is generally conceded that the Warsaw Pact countries have superior conventional forces but that NATO's first-use option (use of nuclear weapons against conventional forces) compensates to some degree. The strategic issue here is that many theorists believe that NATO's nuclear weapons deter aggression, so it does not matter that NATO's conventional forces are weaker than those of the Warsaw Pact. At present NATO has come to rely on the compensatory role of its nuclear forces. It is widely recognized that to change NATO's orientation would be both economically and politically costly. A shift to greater reliance on conventional forces would require much more money and many more troops. NATO countries already have a hard time shouldering the costs they are expected to pay and conscription remains unpopular. Nevertheless, especially in light of the consequences of even a limited tactical nuclear war, these burdens seem preferable to one school in the strategic debate.

Though only briefly, Schmidt argued for such a turn from nuclear reliance. From the strategic community, one can turn to Komer for the argument for restructuring NATO such that it need not rely on tactical nuclear weapons and to Cohen (1975) and Gemeote for the argument that NATO needs a nuclear war fighting emphasis.

More important for the 1970s and 1980s, however, was the Schlesinger doctrine that fueled not only the counterforce but also the NATO debates. Specifically, when asked for examples showing the utility of increased nuclear options, Schlesinger often cited the defense of Europe. Not only did the idea of tactical nuclear war in Europe make Europeans uncomfortable, but they and others also reacted very strongly to one of the new weapons proposed: the neutron bomb. The neutron bomb, billed as killing people rather than property by its opponents, has lowered blast and higher radiation. As a result, its design fits into plans for using tactical nuclear weapons in battlefield situations. To many Europeans, the presence of such weapons makes using nuclear weapons and having a nuclear war much more likely. Within the strategic community, Scott and Cohen (1978) argue for it, and Miettinen (1977) and Kaplan (1978) argue against it. Advocates claim that tactical nuclear weapons deter war and can also be employed in a controlled manner; critics point to the ease with which the line between conventional and nuclear war can be crossed and to the prospect for escalation to full-scale nuclear war if this line is crossed.

The NATO debate soon shifted from tactical to intermediate nuclear weapons with the 1979 NATO decision to deploy 572 American medium-range ballistic and cruise missiles. Gradually, basic strategy has come more into focus, especially with the call for no first use issued by Bundy, Kennan, McNamara, and Smith (Bundy et al., 1982). In other words, an examination of basic strategy is a reconsideration of the move from war-deterring to war-fighting orientations, and in the case of these strategists, a call for at least a return to only war-deterring nuclear arsenals.

Philosophical and Religious Assessments of Deterrence and Strategic Theory

Throughout the nuclear arms race—in fact, throughout the history of warfare—philosophical and religious assessments have been made. These assessments are usually made by persons who are members of neither the strategic community nor the politico-military establishment. Philosophical and religious perspectives are often highly critical of the policies that govern the threat of use and actual use of weapons. Because of the focus on ethical and moral issues, as much as anything, philosophical and religious voices are usually excluded from direct policy-authoritative influence, especially in the nuclear age.[12]

At present, the realist model dominates political science, strategic studies, and official policymaking. The realist model has its philosophical roots in Thomas Hobbes, the early modern British philosopher who argued that sovereign nations stand in a state of nature in relation to one another. A state of nature is the situation that exists when a common authority is lacking. While nations have such authority internally (domestically), they do not have it externally (internationally). For Hobbes, a state of nature is a state of war (potential or actual). The political realist ultimately relies on military strength to guarantee national security.

As a result, and despite the rhetoric used by most national governments, ethical and moral issues are largely irrelevant in international politics: both war and preparation for war are viewed as amoral. In this view, since moral principles can operate only within a framework of common authority, the state of nature existing among nations means their interactions should not be conducted or judged in moral terms. Following Hobbes, the realist perspective sees no efficacy to moral princi-

---

12. There are, of course, exceptions. For example, Thomas Murray (1960), a Roman Catholic and member of the Atomic Energy Commission, did inject a moral influence, and David Singer (1962) tried to combine the work of formal strategists and nuclear pacifists. See Freedman, p. 418, n.21 and 32.

ples or international treaties that cannot be enforced. Since no global authority can enforce such principles and treaties, nations must ultimately defend themselves. Preservation of national sovereignty—whether consistent with moral and legal standards—becomes the bedrock for formulating and executing policy decisions. This realist view is opposed by most philosophers and theologians who are termed idealists by the realists. The philosophers and theologians, in turn, view the realists as advocating immoral positions.

The public is typically more influenced by philosophical and religious reasoning than the political establishment is, and some indirect policy influence has resulted. Since the public more easily understands, more regularly hears of, and is more likely to read philosophical-religious arguments than strategic ones, the literature from this perspective is often erroneously assumed to be dominant by the public even though quantitatively it represents a small portion of the nuclear literature.

In this section, we indicate some of the main topics in the philosophical and religious responses to the nuclear age. Some general criticism of deterrence and strategic theory comes as much from theoretical and personal dissatisfaction with the entire enterprise or with the key facets of it as it does from philosophical or religious orientations. From a theoretical perspective, Green provides a wholesale critique of theory of nuclear deterrence and paints it as a pseudoscience (even though he operates from an explicit ethical position as well). From personal negotiating experience, Myrdal provides a devastating critique of U.S. and Soviet approaches to arms control and suggests their true intent does not square with their statements (even though she too makes explicit her globalist values). Despite the presence of many texts such as these (see also Rapaport, 1960 and 1964), we proceed here for the most part with the narrow focus of those works in which the philosophical or religious perspective is predominant. Moreover, because this literature is so large yet has had so little influence, we cite in our bibliography for this section mostly books, entire journal issues on the topic, and only a few particularly influential or insightful articles.

We address the philosophical and religious assessments in subsections on the following seven topics: (1) normative issues in the nuclear debate; (2) nuclear war and human extinction; (3) just war and religious ethics; (4) deterrence and game theory; (5) strategy and morality; (6) power and domination; and (7) nuclear weapons and western metaphysics.

Normative Issues in the Nuclear Debate

Philosophy and religion are closely associated with norms or values. In addition to the use of logical and conceptual tools by professional philosophers and theologians, normative issues are often focal in their analysis of issues. In terms of the nuclear debate, this normative concern is related to both professional activity and public policy analysis.

In terms of professional activity, the issue concerns the social responsibility of philosophy and religion. Throughout the nuclear age, a few philosophers and theologians have argued that the study of and a response to nuclear issues should be regarded as an appropriate, if not obligatory, professional activity. Since there is a tendency for philosophy to be abstract and for theology to be other-worldly, a call for an application of professional skills to concrete and worldly problems gets issued again and again over the years. Early in the nuclear age, this call came more from British and continental thinkers than from Americans. Because these thinkers are more closely associated with arguments over whether nuclear war will bring about the extinction of the human species, they are treated in the next subsection.

In the United States Dewey and Schilpp in the 1940s and a group of philosophers and others associated with Ginsberg in the 1960s urged that nuclear issues be addressed as a matter of social responsibility. This call has been made even more frequently in the 1980s. Gay (1982) argues that philosophers have a social responsibility in the nuclear age to analyze critically both public and governmental assumptions regarding nuclear war, and Govier addresses several of the areas where philosophers can apply their skills, such as whether it is ethical to work in defense production, the status of nation states, our obligation to future generations, the meaning of life, and problems of self-deception. In other words, they and others argue there are many topics in the nuclear debate to which philosophers and theologians can and should apply their skills.

In terms of public policy analysis, the issue concerns which normative framework to apply. Basically, normative questions can be approached in two ways. On the one hand, one can first work out moral principles and consider actions by themselves (apart from their consequences). This approach is termed deontic or deontological (the study of "oughts") and is done *a priori*, before considering what is true in the world at present and what would follow if these actions were taken. Many advocates of this approach are moral absolutists: the actual situation does not alter what is the moral thing to do, and one should never do what is wrong.

On the other hand, one can also relate moral assessment to the effects of various actions. This approach is termed "consequentialist" and is done *a posteriori* (considering what will likely be the case after an action is taken). There are many consequentialist approaches, such as just war theory, utilitarianism (including game theory), and pragmatism. Arguments rage over which approach is more applicable to nuclear issues and on what version of each approach is correct. An application of an absolutist model can be found in Kenny and in Volbrecht (in Gay, 1984), and a critique of the applicability of the deontic approach can be found in Shaw (1984). For a general discussion of these distinctions and their relation to strategic studies, see Hardin and Mearsheimer (in Hardin et al., 1985). For an illustration of how one can reach negative conclusions concerning war starting from any of these approaches, see Cady (in Gay, 1984), Goodin, and Stevenson.

So much has been written from just war and utilitarian (especially game theory) perspectives that they require separate subsections here. Other issues (traditional strategy, politics, and metaphysics) also capture philosophical attention and are treated in subsequent subsections. Nevertheless, the topic that has perhaps captivated most public and professional attention is the prospect for extinction; it is to this debate that we turn in the next subsection.

Nuclear War and Human Extinction

Throughout the nuclear arms race arguments have been made that nuclear war will lead to human extinction. Various factors are cited, such as fallout, ozone depletion, and nuclear winter (see Chapter 5 for a discussion of these issues). Those who make these arguments usually insist that nuclear war should never be waged and that nuclear weapons should be eliminated as soon as possible. The debate on this topic largely concerns whether extinction is a likely factual consequence and whether such a consequence can or should normatively ground a moral condemnation of nuclear war and nuclear weapons.

This debate has gone through several phases and the contrast in views is quite sharp. In the 1940s and 1950s, most of the voices came from major British and continental philosophers and theologians, and it is fascinating to recall their arguments, which are still being made today in somewhat updated forms.

One of the first to speak out was Pierre Teilhard de Chardin, who in September 1946 wrote "Some Reflections on the Spiritual Repercussions of the Atom Bomb" (see Teilhard de Chardin, 1964). He believed that everything, including nuclear weapons, was a part of a process of "complexification" (while our world gets more complicated, it is progressing toward even more integrated and higher forms). As a result, he does not fear nuclear destruction. Teilhard de Chardin's faith in the emergence of greater and greater consciousness in which no new piece of knowledge, not even that of nuclear weapons, would destroy the direction of the process was not shared by other leading theologians and philosophers.

Among those calling for an immediate end to the nuclear arms race because of their fear for the survival of nations and peoples were Albert Schweitzer (1958) and Bertrand Russell (1959 and 1962). While Schweitzer was the most emphatic in a call for disarmament, Russell stressed the more fundamental need for reduction of East-West tension. In an earlier essay, "The Future of Mankind" (1950), Russell accepted the threat of war against the Soviet Union if it would not negotiate something like the Baruch plan for international control of nuclear weapons. Because of the growth of the Soviet stockpile of nuclear weapons and the development of the hydrogen bomb, Russell rejected this earlier view in one of the appendixes of *Common Sense and Nuclear Warfare* (1959).

Even more curious than the case of Teilhard de Chardin is that of Karl Jaspers (1961). Like Schweitzer and Russell, he presents the disjunction of destruction or fundamental change; unlike them, he sees conditions under which it would be moral to choose destruction. Jaspers argues we should risk destruction in nuclear war if the alternative were losing our "humanity" under totalitarianism.

While there were some early American contributions to this debate (see Palter, 1964), most of the American debate has taken place in the 1980s, particularly since Schell's *The Fate of the Earth* (1982). Schell stresses the catastrophic consequences of nuclear war, suggesting human extinction as a real possibility, and argues for immediate nuclear disarmament as the only moral response to what he presents as the absolute immorality of even taking the risk of human annihilation. While Schell tries to make his argument philosophical, he is not a professional philosopher, and, among philosophers, even a more stark version of his thesis can be found.

Somerville (1983) and Santoni (1984) have coined the term "omnicide" as the irreversible extinction of all sentient life (see also Somerville in Fox and Groarke, and Santoni in Gay, 1984). They argue that nuclear war could destroy all life. Santoni argues, further, that because some nuclear policies express the willingness, under certain conditions, to incapacitate or destroy the adversary nation, they are genocidal in intention and imply "multiple genocide." He appeals to such sources as the Nuremberg tradition and United Nations resolutions against genocide to condemn nuclear war planning and fighting by any government and to demand individual and collective opposition and nonviolent resistance by all citizens to such governmental acts.

Schell, Somerville, and Santoni have their critics. Routley (in Gay, 1984) argues that extinction is not factually likely and that the conclusions drawn are morally problematic. The latter point concerns the assumption that values require persons and that anything that destroys all persons would be wrong because it would destroy all values. He further contends that not everyone is equally threatened or equally responsible and that calls for world government are too hasty. Gay (1982) argues that physically it is possible that nuclear war can have human survivors and societal recovery, and that ethically basing condemnation of nuclear war on worst-case scenarios is not the best approach. In this regard, he notes that if war at low levels is wrong, war at high levels is wrong, and that showing this to be the case is a stronger condemnation of war than that which only condemns the prospect of omnicide.

Routley and Gay are not critical of the call to eliminate nuclear weapons but of the manner in which this argument is made. They would rather concede some ground to the nuclear establishment than undercut their criticism by the use of possibly exaggerated claims about the consequences of nuclear war, and they believe that even on the postattack assumptions of governments, nuclear war and current nuclear policies are immoral. Other contributions to this debate can be found in Aronson, Bordo (in Fox and Groarke, 1985), Gallie (in Blake and Pole, 1983), Soper (in Blake and Pole, 1984), and Shibata.

Just War and Religious Ethics

Two overlapping topics of concern are just war and religious ethics. The theory of just war dates back to the work of Saint Augustine and Saint Thomas. While this tradition affirms that some wars are just, it also lays down the criteria for distinguishing just from unjust war. Christian ethics and other systems of religious morality refer to broader concerns than the criteria for just war that are of relevance to the nuclear debate. For example, pacifists, along with just war theorists, are often found within the ranks of Christians, and both groups find Biblical support for their views. While some just war theorists do not appeal to religious values per se and while Christian ethics include more positions on war than only just war theory, it is hard to separate discussion of these topics when addressing nuclear deterrence and strategy. Here we stress some common themes.

Respect for the distinction between combatants and noncombatants made easy the transition from the critique of conventional to nuclear weapons and strategy. Already in 1944, Ford, in "The Morality of Obliteration Bombing," had used this distinction to argue against the shift to strategic bombing (see Wasserstrom, 1970). Soon after the war, various churches quickly concluded that from a theological and moral perspective, nuclear weapons and nuclear war, especially when they destroyed or even threatened innocent populations, were unacceptable. One who denied that persons have the right to risk God's creation was Gollancz. He affirmed a Christian ethic in which faith in Christ is our security even under the risk of unilateral nuclear disarmament. This argument would be termed idealistic by the realists. The response of Gollancz is "unless the impossibilism of Christ is substituted for the possibilism of politics, the world must destroy itself."[13]

---

13. Victor Gollancz, *The Devil's Repertoire of Nuclear Bombing and the Life of Man* (Garden City, N.Y.: Doubleday, 1959), unnumbered page entitled "A Note for the American Edition."

While the peripheral work of Gollancz illustrates a use of Christian ethics that has affinity with the arguments against human extinction considered in the preceding subsection, the mainstream work of Ramsey (1961, 1963, and 1968) is central to just war theory. One of his more celebrated arguments stems from his claim that it would be wrong, even if it worked, to tie babies to automobile bumpers in order to get people to drive safely. He sees hostage populations in the nuclear age as presenting the same problem of the innocent being immorally threatened. One of his critics, Walzer, argues that nuclear deterrence is not a threat that restrains us or deprives us of our rights, so it is not immoral. See, too, the criticisms of Ramsey's critic O'Brien (1967, 1983).

These just war analyses have not remained abstract. Roszak argued early in the McNamara era that pursuit of either countervalue or counterforce nuclear policy violates just war criteria and that since the United States was pursuing both with MAD and flexible response, its nuclear policies were immoral. Whether just war theory can ever be used to sanction any nuclear options has been questioned by Wells (1969, 1984) and Santoni (1985). Surveys of the historical development and current viability of just war can be found in Churchill and Johnson (1981).

A wide range of just war and religious views has gradually emerged concerning what should be done in light of the fact that nuclear weapons now exist. For example, Christian writers defend positions as diverse as the view that deterrence is morally acceptable and the view that unilateral nuclear disarmament is morally dictated (Goodwin). Some of the early reflections can be found in the Calhoun Commission (1946, 1959) and British Council of Churches (1946, 1959). In fact, most nationally organized religious groups have issued periodic moral assessments of nuclear weapons and politics. Herzog, Levine, and Driver cover many of the groups during the early period, and Potter provides a helpful annotated bibliography on the religious literature of this period. More recently, Wallis has provided an interesting collection of more recent statements by diverse church leaders. For even more recent reflections, see National Conference of Catholic Bishops, Dimler, and Hollenbach. Also relevant are Batchelder, Bennet (1962), Clancy, Gessert, Geyer, Johnson (1984), Nagle, Newman, Stein, and Tucker.

Deterrence and Game Theory

Moral assessments of deterrence are often based on consequentialist approaches. Recently, several philosophers have tried to use game theory to buttress their arguments. In game theory, one sets up a model for conflict situations and assigns numerical values to the possible outcomes and then calculates which set of moves rational players should select. Whereas most formal strategists who used game theory concluded deterrence works, most philosophers who use game theory to assess deterrence conclude it is immoral. The uses of probability and game theory in nuclear strategy are discussed at length in the first section of Chapter 4, and many of the points made there apply here too.

In deciding on nuclear policy, the United States and the Soviet Union can, for example, arm or disarm. For each possible outcome, the game theorist assigns numerical values. Typically, nuclear strategists give a big payoff (for example, plus 10) for the side that arms when the other disarms, and they do not assign very negative numbers to mutual arms racing (say, minus 5) or very positive numbers to mutual disarming (say, plus 5). (See figure 1 in Chapter 4.) As a result, it can appear to be rational (and justified) to continue to base security on deterrence.

In contrast, several philosophers assign quite different values— mutual disarmament gets very high points and superiority gets very low points. As a result, nuclear disarmament, even unilaterally, appears to be rational and justified (moral). This line of reasoning, though of a much more complicated form, can be found in Lackey (1975, 1982, and 1984), Groarke (in Fox and Groarke, 1985), and Measor (in Blake and Pole, 1983). Each uses game theory to reach the conclusion that nuclear disarmament should be pursued. Some philosophers using game theory end up more in the camp of nuclear deterrers. Their assignment of numbers is closer to that of the formal nuclear strategists. This perspective is illustrated in the work of Gauthier, Kavka (1978, 1980), and Hardin. See also MacLean and Paul.

Both sets of conclusions by philosophers using game theory to assess nuclear policy share the assumption that game theory is applicable in this context. As noted in Chapter 4, Rapaport and Green argued two decades ago that the use of game theory in planning nuclear strategy is pseudoscientific and dangerous. Among philosophers, Wolff (1970) made a similar argument in his seminal essay on the topic. More recently, Benn has shown some of the shortcomings in the philosophical use of game theory to assess nuclear policy.

Strategy and Morality

Some efforts have been made to bring together, at least in print, members of the strategic and philosophical-religious communities. These efforts reveal that most strategists are realists and most philosophers are idealists. Perhaps the most that occurs is clarifying one's own perspective and understanding the other's perspective. Little in the way of change or synthesis emerges. Nevertheless, those efforts at least thematize the differences and represent a start toward dialogue between the two camps.

Kaplan (1973) was one of the first to bring together strategic and moral analyses. The collection by Ford and Winters has both sympathetic and unsympathetic contributions, but the editors' own ethical assessment is that the Schlesinger doctrine is unacceptable. Their book, which focuses on counterforce options, anticipates many more recent analyses. In the special issue of *Ethics* on ethics and nuclear deterrence (Hardin et al., 1985) the articles are about equally divided between representatives of various realist schools in strategic theory and various analytic schools in Anglo-American philosophy. Almost all the articles focus on deterrence, but none come from just war or continental perspectives. While the special issue of *Ethics* resulted from a face-to-face encounter between strategists and philosophers, Sterba offers a contrived but still useful confrontation between philosophical and nonphilosophical sources. His collection covers more themes and has representatives from more philosophical traditions. Cohen and Lee also have a collection of philosophical and nonphilosophical essays.

Power and Domination

Although the philosophical and religious literature on the nuclear debate is often presumed to be composed only of moral assessments, much of it is in fact more properly classified as political philosophy. In political philosophy, the focus is on the nature of the state—whether a nation is legitimate, which forms of government are preferable, and how nations should act toward their own citizens and toward other nations. In the philosophical and religious literature on the nuclear debate, political assessment concerns issues of power and domination. More specifically, the focus is usually on whether *Realpolitik* is viable in the nuclear age, whether nuclear weapons render the nation state obsolete, and how an understanding of the Soviet and American systems affects the assessment of nuclear policies.

Although most members of the strategic community accept *Realpolitik*, only a few of the philosophers and theologians who address nuclear issues operate from Hobbesian assumptions. One of these is Morris (in Hardin et al.), who accepts the Hobbesian premise that once one nation attacks another, nations are in a state of nature (the state of war) in which principles of morality (noncombatant immunity) are no longer in effect. He argues, on this basis, that nuclear deterrence is an appropriate means for pursuing national security. Among the critics of *Realpolitik* are philosophers such as Brunk (in Fox and Groarke, 1985), Donaghy (in Gay, 1984), and political scientists such as Beres (in Gay, 1984). These critics bring out the Hobbesian assumptions and argue that in the nuclear age these premises do not convey realism. The maximization of force (building massive nuclear arsenals) and distrust (the disinterest in significant arms control treaties) is the formula for destruction, not security. Out of this critique, the conclusion is usually reached that, in addition to pursuing major reductions in nuclear weapons, nations should move toward a workable form of global security.

The thesis of global security is often debated in terms of whether it is compatible with the maintenance of the traditional concept of national sovereignty. Schell (1982) is perhaps most widely associated with the view that elimination of nuclear weapons and maintenance of the nation state are incompatible, though he subsequently modified his view (Schell, 1984). Most philosophers, however, are somewhat hesitant to propose world government as the solution. Most advocate deep cuts in weapons and the pursuit of less violent and more lawful means for resolving international security. One of the more thoughtful essays in this regard is by Woodward (in Fox and Groarke, 1985). See too Geyer and Narveson (in Fox and Groarke, 1985).

The topic of the Soviet and American systems brings out most clearly the issues of power and domination. Those who make arguments in these terms view most of the ethical assessments as too abstract. Moral judgment, they generally argue, should not be separated from historical analysis and political assessment of the Soviet and American systems. For some the Soviet Union is the "demonic enemy" and must be inhibited in its urge to dominate, even if this effort leads to nuclear war. Among American philosophers, this view can be traced back at least to Sidney Hook. Shortly after Eugene Burdick's popular novel *Fail Safe* gave the impression that we need primarily fear loss of control over nuclear forces and accidental nuclear war, Hook took to task this type of thinking. He was less concerned about accidental nuclear war than intentional Soviet aggression.

This view of the Soviets continues and is dominant among the contributors to Hardin et al. Increasingly, political philosophers, especially those trained in the continental traditions, are challenging this characterization of the Soviets. Howard and Anderson (in Gay, 1984) argue that the actions of both nations need to be put in the proper historical context and that the long-standing global problem of domination needs to receive more attention—solidarity among peoples, regardless of the nations in which they live, should be our goal. On this topic, Somerville has perhaps made the largest contribution. On the one hand, he questions the compatibility of U.S. nuclear policy with the principles of American democracy (Somerville, 1975, 1978). On the other hand, he shows that the principles of Marxism are consistent with the aims of the peace movement in the nuclear age and that the Soviet Union can renounce war (Somerville, 1976, 1981).

Nuclear Weapons and Western Metaphysics

Philosophers and theologians are often characterized as asking "What is the meaning of life?" and "What is the basic nature of the universe?" These questions are treated by metaphysics, which addresses first principles or basic assumptions. While metaphysical questions can be asked about such general topics as life and the universe, they can also be asked about much narrower and more specific topics—such as "What is the meaning of nuclear weapons?" and "What is the nature of the relation of nuclear states to the world and its inhabitants?" The answers to these questions tell us a lot about what we have become and what we can be. Most of these metaphysical analyses are influenced by Kant, Hegel, Heidegger, or Eastern thought.

Nuclear weapons themselves have received some attention. While Shrader-Frechette (in Fox and Groarke, 1985) explores the philosophical links between nuclear reactors and nuclear weapons, it is Weinberger and Kilbourne (in Gay, 1985) who take the analysis to a deeper metaphysical level using Heidegger's comments on death and technology as a point of departure.

Weinberger notes how nuclear weapons alter the past concept of death. We no longer presume that after our death life will go on—we may be subjected to a collective death. This possibility negatively affects our metaphysical self-understanding, in that our sense of world has been altered. Kilbourne brings out how our desire to control nature and persons has gotten us to a point with nuclear weapons that we look more to technological capability of weapons systems than to policy statements of nation states in our efforts to understand one another.

Those who turn to metaphysics in analyzing the nuclear age are concerned about the conceptual choices made in the West regarding how to relate to nature and persons, in particular, our choice to have dominion over or to control both. Some turn to Kant. Schilpp and Friedrich (in Ginsberg, 1969), and Hampsch (in Gay, 1985) place hope in Kant's idea of perpetual peace being achieved by a federation of nations. Others, such as Butler (in Gay, 1985), use Hegel to argue that peace is contingent (there is no guarantee war can be eliminated as a human possibility) and that the real problem lies much deeper than the history of nation states, particularly such fairly recent ones as the United States and the Soviet Union. Basically, the world has not moved beyond the master-slave dialectic, or the problem of domination. As a result, we now must pass through nuclear trauma either in war or in imagination and then move to new forms of social interaction.

Perhaps the deepest penetration into our plight has been offered by those who combine Heidegger's probing questions with Eastern alternatives. As early as 1962, Joseph Schorstein turned to Heidegger to present the atomic bomb as one of the last and most tragic consequences of the West's loss of "nearness" to nature and persons. More recently, Zimmerman (1983, and in Fox and Groarke, 1985), Heim, Litke (in Fox and Groarke, 1985), and Dombrowski either have used Heidegger to thematize the West's overstress on control or have turned to Eastern traditions for styles of relating to nature and persons which are based more on harmony than exploitation. For a comprehensive review of the literature on Heidegger and nuclear weapons, see Smithka.

The nuclear debate can lead to philosophical reflection that calls into question the basic self-understanding of Western culture. The result of this culture appears to be nihilistic, especially in the nuclear age: we may literally go from being to nothing. But, just as Heidegger showed the disastrous consequences of Western metaphysics, even so he pointed to a new path, more Eastern in style, which may yet be our salvation. He calls for a new style of thinking that is more meditative than calculative, more receptive than aggressive. Heidegger calls on us to ''let beings be.'' We can live in peace if we allow it. We can be open to life rather than death and create a clearing for being rather than the space of nothing.

Summary

In this chapter we have focused on defenses and criticisms of deterrence and have treated separately the arguments by members of the strategic community and the arguments of those outside the community. The first three sections cover the very large and influential literature produced by the strategic community. These sections reveal that throughout the nuclear age, two basic factions have been present—the nuclear deterrers and the nuclear warriors. We noted how neither faction had much influence on policy formulation during the 1940s and 1950s, how the nuclear deterrers came to shape policy in the 1960s, and how during the 1970s and 1980s both groups have vied to sway policy, each achieving partial influence.

In sharp contrast, the fourth and final section of this chapter treats the assessments of nuclear weapons that came from outside the strategic community. On the one hand, this literature is much smaller in size and has exerted much less influence on the formulation of policy. On the other hand, this literature is predominantly critical. Whereas almost all members of the strategic community see numerous roles for nuclear weapons, those writing from outside the strategic community find few, if any, legitimate and moral roles for nuclear weapons. What both groups have in common is that they focus on the role of nuclear weapons, whether their assessment is positive or negative. In Chapter 8, we take up these groups again, but examine instead their respective visions for the future. We look at arguments that: (1) we must rely on nuclear weapons, (2) we can get rid of nuclear weapons, and (3) we can have a world without weapons and war.

Bibliography

The following bibliography includes references cited in the text and related suggested readings; references are organized according to the major divisions of the chapter. The full bibliographic reference for all sources cited in the text of this chapter may be found in either the annotated or unannotated bibliography corresponding to the division of the text in which the source is cited. Sources that are annotated are the most essential or significant.

The Independence of Policymaking from the Strategic Community: 1945 to 1960

Borden, William L. *There Will Be No Time.* New York: Macmillan, 1946.

One of the earliest works on nuclear strategy. This, until recently, largely neglected book was influenced by Brodie's *The Absolute Weapon.* Borden's counterforce scenarios and speculations on development and use of missiles demonstrate the degree to which he anticipated much of the subsequent development of nuclear weapons systems and policies.

Brodie, Bernard ed. *The Absolute Weapon: Atomic Power and World Order.* New York: Harcourt, 1946.

This early and very influential book on nuclear strategy is closely linked with the perspective of Brodie, the book's editor and main contributor. (It also includes pieces by Fredericks Dunn, Arnold Wolfers, Percy E. Corbett, and William T. R. Fox.) The book actually argues that nuclear weapons (then only the atomic bomb delivered by aircraft) are not absolute in the sense of preventing or winning war. The book argues for limited use (largely counterforce to avoid destruction of population). Among other things, Brodie cites and details eight ways nuclear weapons have changed war.

Viner, Jacob. "The Implications of the Atomic Bomb for International Relations." *Proceedings of the American Philosophical Society* 20 (January 1946): 53–58.

Often cited as the first contribution to nuclear strategic theory. Viner, an economist at the University of Chicago, delivered this paper on November 16, 1945, at the Symposium on Atomic Energy and Its Implications, a joint meeting of the American Philosophical Society and the National Academy of Sciences. Both the symposium and Viner's paper are important as initial academic and strategic assessments, respectively. Viner makes countervalue/counterforce and first-strike/second-strike distinctions, notes the lack of defense, and foresees the end of victory. He rejects the feasibility of world government, supports the United Nations, and encourages "the conscientious and unrelenting practice by the statesmen of the Great Powers, day after day, year after year, of mutually conciliatory diplomacy" (p. 57).

Blackett, Patrick M. S. *The Military and Political Consequences of Atomic Energy.* London: Turnstile, 1948.

Brodie, Bernard. "Nuclear Weapons: Strategic or Tactical?" *Foreign Affairs* 32 (January 1954): 217–29.

————. "Strategy as a Science." *World Politics* (July 1949).

Bush, Vannevar. *Modern Arms and Free Men.* New York: Simon & Schuster, 1949.

LeMay, Curtis E. with MacKinley Kantor. *Mission with LeMay: My Story.* Garden City, N.Y.: Doubleday, 1965.
U.S. Congress, House Committee on Armed Services. *Investigation of the B-36 Bomber Program.* Washington, D.C.: Govt. Print. Off., 1949.
_____. Senate Committees on Armed Services and Foreign Relations. *Military Situation in the Far East, Hearings.* Washington, D.C.: Govt. Print. Off., 1951.
Wohlstetter, Albert J., *et al. Selection and Use of Strategic Air Bases.* Rand Report R-266. Santa Monica, Calif.: Rand Corp., April 1954.

The Influence of the War-Deterring Strategic School on Policymaking: 1960s

Brennan, Donald G., ed. *Arms Control, Disarmament, and National Security.* New York: Braziller, 1961.

Very much based on the issue of *Daedalus* (see below) devoted to arms control. This book has twenty-three authors. Eleven of the articles are essentially the same as they appeared in *Daedalus,* five have moderate to extensive revisions, and five are new pieces.

Frisch, David H., ed. *Arms Reduction: Programs and Issues.* New York: Twentieth-Century, 1961.

From the work of physical and social scientists working at MIT in the summer of 1960, and conducted under the auspices of the American Academy of Arts and Sciences. Thomas Schelling and Morton Halperin's *Strategy and Arms Control* (see below) is a companion volume. The MIT group was headed by Bernard Feld, and the book has contributions by such figures as Feld, David H. Frisch, Louis B. Sohn and Joseph Salerno.

Holton, Gerald, ed. "Arms Control." *Daedalus* 89 (Fall 1960).

Journal of the American Academy of Arts and Sciences devoted this issue to arms control. Jerome Wiesner wrote the foreword to show that a theory of arms control was central to the strategy of the war-deterring school of the strategic community. Donald Brennan, Robert Bowie, and William R. Frye write on the background to arms control; Herman Kahn, Edward Teller, Henry Kissinger, Paul Doty, A. Doak Barnett, and Kenneth E. Boulding write on major issues and problems; Bernard Feld, Louis Sohn, Thomas Schelling and Jerome Wiesner write on implementation; Saville Davis, Hubert H. Humphrey, Ithiel Pool, William Fox, and Erich Fromm write on formation of U.S. arms control policy; Harrison Brown and Arthur Larson write on beyond the cold war; and Christopher Wright provides a helpful selective critical bibliography. Most of these articles, plus some new ones, were subsequently published, in edited form, as a book (see Brennan above).

Kaufmann, William W., ed. *Military Policy and National Security.* Princeton, N.J.: Princeton Univ. Pr., 1956.

Considered by many to be the hallmark of the war-deterring or liberal school in strategic studies, this work arose as a critical response to strategy of massive retaliation, associated with Princeton's Center of International Studies. Kaufman cites his indebtedness to Bernard Brodie. Contributors include William Kaufmann, Roger Hilsman, Klaus Knorr, and Gordon Craig. Kaufmann later influences policy of McNamara.

Schelling, Thomas C. and Morton H. Halperin. *Strategy and Arms Control.* New York: Twentieth Century Fund, 1961.

Companion volume to Frisch (see above) and prepared under the same conditions. Treats arms control in relation to general and limited war and the arms race itself. Has chapters evaluating arms control proposals, as well as a chapter that suggests how to make arms control work.

Stockholm International Peace Research Institute. *Arms Control: A Survey and Appraisal of Multilateral Agreements.* New York: Crane, Russak, 1978.

This work is by a nongovernment research institute, focusing on multilateral (rather than U.S.-Soviet) agreements, and covers more than just nuclear weapons.

Amster, Warren. "Design for Deterrence." *Bulletin of the Atomic Scientists* 11 (May 1956): 164–65.

————. *A Theory for the Design of a Deterrent Air Weapons System.* San Diego, Calif.: Convair Corp., 1955.

Arms Control: Readings from "Scientific American." San Francisco: Freeman, 1973.

Bechhoefer, Bernard. *Postwar Negotiations for Arms Control.* Washington, D.C.: Brookings, 1961.

Brodie, Bernard. *Escalation and the Nuclear Option.* Princeton, N.J.: Princeton Univ. Pr., 1966.

————. *Strategy in the Missile Age.* Princeton, N.J.: Princeton Univ. Pr., 1959.

Bull, Hedley. *The Control of the Arms Race: Disarmament and Arms Control in the Missile Age.* London: Weidenfeld & Nicolson, 1961.

Burns, Richard Dean. *Arms Control and Disarmament: A Bibliography.* Santa Barbara, Calif.: ABC-Clio, 1977.

Buzzard, Anthony, et al. *On Limiting Atomic War.* London: Royal Institute of International Affairs, 1956.

Chayes, Abram and Jerome B. Wiesner. *ABM: An Evaluation of the Decision to Deploy an Antiballistic Missile System.* New York: New Amer. Lib., 1969.

Dentler, Robert A. and Phillips Cutright. *Hostage America: Human Aspects of a Nuclear Attack and a Program of Prevention.* Boston: Beacon, 1963.

Halperin, Morton H. *A Proposal for a Ban on Use of Nuclear Weapons.* Washington, D.C.: n.p., 1961.

Henkin, Louis. *Arms Control: Issues for the Public.* Englewood Cliffs, N.J.: Prentice-Hall, 1961.

Hoag, Malcolm W. "NATO: Deterrent or Shield." *Foreign Affairs* 36 (January 1958): 278–92.

Kahn, Herman. *On Thermonuclear War.* Princeton, N.J.: Princeton Univ. Pr., 1960.

Kincade, William and Jeffrey D. Porro, eds. *Negotiating Security: An Arms Control Reader.* Washington, D.C.: Carnegie Endowment for International Peace, 1979.

King, James E., Jr. "Nuclear Plenty and Limited War." *Foreign Affairs* 35 (January 1957): 238–56.

Kissinger, Henry. *The Necessity for Choice: Prospects of American Foreign Policy.* New York: Anchor, 1960.

————. *Nuclear Weapons and Foreign Policy.* New York: Harper, 1957.

Lapp, Ralph. *Kill and Overkill.* New York: Basic Books, 1961.

Liddell Hart, Basil. *Deterrent and Defense: A Fresh Look at the West's Military Position.* New York: Turnstile, 1960.

Nitze, Paul. "Atoms, Strategy and Policy." *Foreign Affairs* 34 (January 1956): 187–198.

Osgood, Robert Endicott. *Limited War: The Challenge to American Strategy.* Chicago: Univ. of Chicago Pr., 1957.

Ranger, Robert J. *Arms and Politics, 1958–1978: Arms Control in a Changing Political Context.* Toronto: Macmillan of Canada, 1979.

Read, Thornton. *A Proposal to Neutralize Nuclear Weapons.* Princeton, N.J.: n.p., 1960.

Schelling, Thomas. *Strategy of Conflict.* New York: Oxford Univ. Pr., 1960.

Sherwin, C. W. "Securing Peace through Military Technology." *Bulletin of the Atomic Scientists* 12 (May 1956): 159–64.

Singer, David. *Deterrence, Arms Control, and Disarmament.* Columbus, Ohio: Ohio State Univ. Pr., 1962.

Smith, Gerard. *Doubletalk: The Story of the First Strategic Arms Limitations Talks.* New York: Doubleday, 1980.

Snyder, Glenn. *Deterrence and Defense.* Princeton, N.J.: Princeton Univ. Pr., 1961.

Stockholm International Peace Research Institute. *Arms Control: A Survey and Appraisal of Multilateral Agreements.* New York: Crane, Russak, 1978.

Tucker, Robert C., et al. *Proposal for No-First Use of Nuclear Weapons: Pros and Cons.* Princeton, N.J.: Center for International Studies, 1963.

United Nations. *The United Nations and Disarmament, 1945–1970.* New York: United Nations, 1970.

————. *The United Nations and Disarmament, 1970–1975.* New York: United Nations, 1976.

————. *The United Nations Disarmament Yearbook.* New York: United Nations, 1976- .

U.S. Arms Control and Disarmament Agency. *Arms Control and Disarmament Agreements, Texts and History of Negotiations.* Washington, D.C.: Govt. Print. Off., 1980.

Wohlstetter, Albert. "The Delicate Balance of Terror." *Foreign Affairs* 37 (January 1959): 211–34.

————, et al. *Protecting U.S. Power to Strike Back in the 1950s and 1960s.* Rand R-290. Santa Monica, Calif.: Rand Corp., September 1956.

Yanarella, Ernest. *The Missile Defense Controversy: Strategy, Technology and Politics, 1955–1972.* Lexington, Ky.: Univ. Pr. of Kentucky, 1977.

The Debate between and Policy Influence of the War-Deterring and War-Fighting Strategic Schools: 1970s and 1980s

Beres, Louis Rene. *Mimicking Sisyphus: America's Countervailing Nuclear Strategy.* Lexington, Mass.: Heath, 1983.

A detailed critique of U.S. nuclear policy of a countervailing strategy. From PD-59 under Carter through the Reagan administration, the influence of the war-fighting strategic school on U.S. nuclear policy is traced. Beres attacks the newest version of limited nuclear war theory, especially its assumption that the threat of counterforce reprisals would improve deterrence. Beres also criticizes the civil defense optimism and arms control disdain characteristic of this policy and strategic theory. Questioning that nuclear deterrence can work indefinitely, Beres argues for a comprehensive test ban, a no-first-use pledge, a nuclear freeze, and nuclear free zones.

_____. "Presidential Directive 59: A Critical Assessment." *Parameters* 11 (March 1981): 19–28.

Beres presents PD-59 as a rejection of MAD, which he considers to be fact, whether policy. He doubts a limited Soviet first strike is likely and that the Soviet would be better deterred if the United States had a similar capability. He stresses that controlled nuclear conflict may be unlikely and targeting of Soviet leadership may further undermine prospect for control. He sees PD-59 as making more likely both a Soviet first strike and accidental nuclear war. Desiring a return to minimum deterrence, he argues for educating the public on the consequences of nuclear war, for governments setting a comprehensive agenda for long-term international security, and for increasing conventional forces.

Brodie, Bernard. "The Development of Nuclear Strategy." *International Security* 2 (Spring 1978): 78–83.

Brodie's last essay. He does a good summary of strategic theory beginning with his own remarks in 1945. As a representative of the war-deterring school, he stresses how well the United States meets the requirements for deterrence. He criticizes the views of such war-fighting strategists as Richard Pipes and Paul Nitze and such correlative policy declarations as the Schlesinger doctrine. He makes the distinctive argument that to increase the president's nuclear options may not be a good idea; in fact, it could do more harm than good and counters constitutional principles concerning warmaking powers.

Bundy, McGeorge, *et al.* "Nuclear Weapons and the Atlantic Alliance." *Foreign Affairs* 60 (Spring 1982): 753–68.

Frequently cited article by some of the leading designers of past U.S. nuclear policy in which the authors (McGeorge Bundy, George Kennan, Robert McNamara, and Gerard Smith), all swayed by the war-deterring school of strategic theory, argue that NATO should renounce its first-use option.

Gray, Colin. "Nuclear Strategy: A Case for a Theory of Victory." *International Security* 4 (Summer 1979).

Gray argues for U.S., but not Soviet, general staff to be able to offer political leadership a plausible theory of military victory. He attacks MAD as policy and the strategic theory of Brodie and other members of the war-deterring school. Wanting a more war-fighting policy than either the Schlesinger doctrine or PD-59, Gray argues for the targeting of Soviet recovery economy and the Soviet State. He indicates that superiority is needed for stability and influence, and he wants the United States to clearly be in this position.

_____. *Strategic Studies and Public Policy: The American Experience.* Lexington, Ky.: Univ. Pr. of Kentucky, 1982.

A historical overview and theoretical assessment of strategic theory and policy by one of the leading members of the war-fighting school. Gray characterizes the school that influenced U.S. nuclear policy during the 1960s as the liberal, arms controller subcommunity and his own school as revisionist and defense minded.

_____ and Keith Payne. *"Victory Is Possible." Foreign Policy* 39 (Summer 1980): 14–27.

Frequently cited article that challenged MAD as a policy. Using criteria of just war theory, Gray and Payne argue U.S. nuclear policy is immoral and suggest the moral solution lies in development of weapons and policies that will make plausible victory in nuclear war at any level. They claim that the distinction between winning and losing is never trivial. They argue against the self-deterrence of MAD; hence, they reject parity and seek clear U.S. nuclear superiority. They do see the potential loss of 100 million Americans to be too high a cost, but they can foresee conditions under which the loss of 20 million Americans could be an option.

Kegley, Charles W., Jr., and Eugene R. Wittkopf. *The Nuclear Reader: Strategy, Weapons, War.* New York: St. Martin's, 1985.

Useful collection of readings. Includes essays by people like Theodore Draper, Albert Wohlstetter, Robert Jarvis, Freman Dyson, Robert McNamara, Kenneth Payne, Colin Gray, Jonathan Schell, Desmond Ball and Carl Sagan. Addresses the war-deterring (MAD) vs. war-fighting (NUTS) controversy. Has moral assessments, including that of the Catholic bishops. Has pro and con selections on ballistic missile defense (star wars). Also addresses civil defense and nuclear winter.

Miller, Steven E., ed. *Strategy and Nuclear Deterrence: An "International Security" Reader.* Princeton, N.J.: Princeton Univ. Pr., 1984.

Collection of influential articles in *International Security* between 1978 and 1983. Includes "The Development of Nuclear Strategy" (see Brodie, 1978) and "Nuclear Strategy: A Case for a Theory of Victory" (see Gray, 1979). Also has Rosenberg's "The Origins of Overkill" (discussed in Chapter 4). Pro and con articles on countervailing strategy.

Thompson, W. Scott., ed. *National Security in the 1980s: From Weakness to Strength.* San Francisco: Institute for Contemporary Studies, 1980.

Key anthology for peace through strength and war-fighting school of strategic theory. Sections on "The Politics of Weakness," "Quick Fixes," "The Politics of Strength," and "An American Strategy for the 1980s." The nineteen selections come from such persons as Elmo Zumwalt, Jr., Fred Ikle, William van Cleave, Richard Burt, Albert Wohlstetter, Edward Luttwak, Kenneth Adelman, Sam Nunn, and Paul Nitze.

Ball, Desmond. *Deja Vu: The Return to Counterforce in the Nixon Administration.* [Los Angeles?] Calif.: Seminar on Arms Control and Foreign Policy, December 1974.

_____. *Developments in U.S. Nuclear Policy under the Carter Administration.* [Los Angeles?] Calif.: Seminar on Arms Control and Foreign Policy, 1980.

Bell, Coral. *The Conventions of Crisis: A Study in Diplomatic Management.* London: Oxford Univ. Pr., 1971.

Bennett, W. S., R. R. Sandoval, and R. G. Shreffler. "A Credible Nuclear-Emphasis Defense for NATO." *Orbis* 17 (Summer 1973): 463–79.

Blackman, Barry M., ed. *Rethinking the U.S. Strategic Posture.* Cambridge, Mass.: Ballinger, 1982.

Bundy, McGeorge. "To Cap the Volcano." *Foreign Affairs* 48 (October 1969).

Canby, Steven. *The Alliance and Europe: Part 4: Military Doctrine and Technology.* Adelphi Paper no. 109. London: International Institute for Strategic Studies, Winter 1974/75.

Carter, Barry. "Nuclear Strategy and Nuclear Weapons." *Scientific American* 230 (May 1974).

Cohen, Samuel T. "Enhanced Radiation Weapons: Setting the Record Straight." *Strategic Review* (Winter 1978).

———— and William C. Lyons. "A Comparison of U.S.-Allied and Soviet Tactical Nuclear Force Capabilities and Policies." *Orbis* 19 (Spring 1975): 79–92.

Davis, Lynn Etheridge. *Limited Nuclear Options: Deterrence and the New American Doctrine.* London: International Institute for Strategic Studies, 1976.

Defending America: Toward a New Role in the Post-Detente World. New York: Basic Books, 1977.

Douglas, Joseph D. "Soviet Nuclear Strategy in Europe: A Selective Targeting Doctrine?" *Strategic Review* 5 (Fall, 1977): 19–32.

Enthoven, Alain and K. Wayne Smith. *How Much Is Enough? Shaping the Defense Program 1961–1969.* New York: Harper, 1971.

Fisher, Roger. *International Conflict for Beginners.* New York: Harper Colophon, 1970.

Geneste, Marc E. "The City Walls: A Credible Defense Doctrine for the West." *Orbis* 19 (Summer 1975): 477–90.

Gray, Colin. *The Future of Land-Based Missile Forces.* London: International Institute for Strategic Studies, 1978.

————. "Presidential Directive 59: Flawed but Useful." *Parameters* 11 (March 1981): 29–37.

————. *The Soviet-American Arms Race.* Farnborough, Hants: Saxon House, 1976.

Greenwood, Ted and Michael Nacht. "The New Nuclear Debate: Sense or Nonsense." *Foreign Affairs* 52 (July 1974): 761–80.

Goure, Leon, Foy Kohler, and Mose L. Harvey. *The Role of Nuclear Forces in Current Soviet Strategy.* Miami: Center for Advanced Studies, Univ. of Miami, 1974.

Hermann, Charles F., ed. *International Crises: Insights from Behavioral Research.* New York: Free Pr., 1972.

Hillenbrand, Martin J. "NATO and Western Security in an Era of Transition." *International Security* 2 (Fall 1977): 3–24.

Holst, Johan and Uwe Nerlich. *Beyond Nuclear Deterrence.* London: Macdonald & Janec, 1978.

Holsti, Ole R. *Crisis, Escalation, War.* Montreal: McGill-Queen's Univ. Pr., 1972.

Ikle, Fred. "Can Nuclear Deterrence Last over the Century?" *Foreign Affairs* 51 (January 1973).

Kaiser, Karl, *et al.* "Nuclear Weapons and the Preservation of Peace: A Response to an American Proposal for Renouncing the First Use of Nuclear Weapons." *Foreign Affairs* 60, no. 5 (Summer 1982): 1157–70.

Kaplan, Fred M. "Enhanced Radiation Weapons." *Scientific American* 238 (May 1978).

Komer, Robert. "Treating NATO's Self-Inflicted Wound." *Foreign Policy* 13 (Winter 1973–74): 34–48.

Lebow, Richard N. *Between Peace and War: The Nature of International Crisis.* Baltimore: Johns Hopkins Univ. Pr., 1981.

Miettinen, Jorma K. "Enhanced Radiation Warfare." *Bulletin of the Atomic Scientists* 33 (September 1977): 32–37.

Moulton, Harland B. *From Superiority to Parity: The United States and the Strategic Arms Race, 1961–1971.* Westport, Conn.: Greenwood Press, 1973.

Nitze, Paul. "Assuring Strategic Stability in an Era of Detente." *Foreign Affairs* 54 (January 1976).

_____. "Deterring Our Deterrent." *Foreign Policy* 25 (Winter 1976–1977).

Osgood, Robert E. and Robert W. Tucker. *Force, Order and Justice.* Baltimore: Johns Hopkins Univ. Pr., 1967.

Panofsky, Wolfgang. "The Mutual-Hostage Relationship between America and Russia." *Foreign Affairs* 56 (October 1973).

Pranger, Robert J. and Roger P. Labrie, eds. *Nuclear Strategy and National Security: Points of View.* Washington, D.C.: American Enterprise Institute, 1977.

Rathjens, George. *The Future of the Strategic Arms Race: Options for the 1970s.* New York: Carnegie Endowment for International Peace, 1964.

Schmidt, Helmut. *Defense or Retaliation.* New York: Praeger, 1972.

Scott, John F. "The Neutron Weapon and NATO Strategy." *Parameters* 7 (1977): 33–38.

Scoville, Herbert, Jr. "Flexible Madness." *Foreign Policy* 14 (Spring 1974).

_____. *Towards a Strategic Arms Limitation Agreement.* New York: Carnegie Endowment for International Peace, 1970.

Smoke, Richard. *War: Controlling Escalation.* Cambridge, Mass.: Harvard Univ. Pr., 1977.

Steinbruner, John. "Beyond Rational Deterrence: The Struggle for New Conceptions." *World Politics* 28 (January 1976).

Stone, Jeremy. *Strategic Persuasions: Arms Limitations through Dialogue.* New York: Columbia Univ. Pr., 1967.

U.S. Congress. Senate Foreign Relations Committee. *Briefing on Counterforce Attacks.* Washington, D.C.: Govt. Print. Off., 1974.

_____. *U.S. and Soviet Strategic Policies.* Washington, D.C.: Govt. Print. Off., 1974.

Van Cleave, William R. and Roger W. Burnett. "Strategic Adaptability." *Orbis* 18 (Autumn 1974): 655–76.

Williams, Phil. *Crisis Management: Confrontation and Diplomacy in the Nuclear Age.* London: Croom, Helm, 1976.

Wohlstetter, Albert. "Is There a Strategic Arms Race?" *Foreign Policy* 15 (Summer 1974).

_____. "Optimal Ways to Confuse Ourselves." *Foreign Policy* 20 (Autumn 1975).

_____. "Rivals but No 'Race'." *Foreign Policy* 16 (Autumn 1974).

Worner, Manfred. "NATO Defenses and Tactical Nuclear Weapons." *Strategic Review* 5 (Fall 1977): 11–18.

York, Herbert. "Reducing the Overkill." *Survival* 16 (March/April 1974).

Young, Oran R. *The Politics of Force: Bargaining during International Crises.* Princeton, N.J.: Princeton Univ. Pr., 1968.

Philosophical and Religious Assessments of Deterrence and Strategic Theory

Blake, Nigel and Kay Pole, eds. *Danger of Deterrence: Philosophers on Nuclear Strategy.* London: Routledge & Paul, 1983.

Seven critical essays on nuclear deterrence mostly by British philosophers. Advocates unilateralism as the best route to a politically independent Europe.

_____. *Objections to Nuclear Defense: Philosophers on Deterrence.* Routledge & Paul, 1984.

Ten critical essays mostly by British philosophers. Authors give moral and political objections to nuclear policies of current Western governments.

Cohen, Avner and Steven Lee, eds. *Nuclear Weapons and the Future of Humanity.* Totowa, N.J.: Rowman & Allenheld, 1984.

Twenty articles, most by philosophers. Articles touch on various themes, especially moral questions concerning deterrence and the consequences of nuclear war.

Eatherly, Claude and Gunther Anders. *Burning Conscience: The Case of the Hiroshima Pilot, Claude Eatherly, Told in His Letters to Gunther Anders, with a Postscript for American Readers by Anders.* New York: Monthly Review Pr., 1962.

Includes preface by Bertrand Russell, foreword by Roger Jungk, and the essay "Commandments in the Atomic Age" by the Vienna philosopher Gunther Anders. While organized around Anders' correspondence with Eatherly, this book functions as an example of applied philosophy: Anders, a moralist, offers practical ethical (and personal) advice to Eatherly on the best way to deal with his predicament. Eatherly, who had been involved with the atomic bombing at Hiroshima, went through much subsequent psychological turmoil, here termed "delayed action effect," that is, later moral reservations. Also, an interesting commentary on public and military reactions.

Ford, Harold P. and Francis X. Winters, eds. *Ethics and Nuclear Strategy.* Maryknoll, N.Y.: Orbis, 1977.

This anthology arose from eight seminar sessions in 1974 to 1975 at Georgetown University, conducted under the auspices of the Woodstock Theological Center and the Institute for the Study of Ethics and International

Affairs. The book reflects the diverse views of strategists and ethicists. The book is specifically designed "to clarify and to probe the ethical consequences of the stark new strategic weapons doctrines which Secretary of Defense James Schlesinger suddenly began emphasizing publicly in early 1974" (p. 1). Editors find the Schlesinger doctrine to be incompatible with proper ethical principles. In addition to a good glossary on nuclear weapons and arms control terminology, the book has two excellent bibliographies, one on politico-military issues and the other on ethical issues.

Fox, Michael Allen and Leo Groarke, eds. *Nuclear War: Philosophical Perspectives.* New York: Peter Lang, 1985.

Philosophers from diverse traditions provide twelve articles and fourteen commentaries, with follow-up sections on "Issues to Think About and Discuss" for each of the five parts: (1) nuclear delusions, (2) the individual and the state, (3) the environment, (4) conceptual and psychological dilemmas, and (5) the pursuit of peace. Uses tools of philosophy for "the task of analyzing and breaking down some of the barriers to clear thinking and ultimately to peace."

Gay, William C., ed. *Philosophy and Social Criticism* 10 (Winter, 1984): "Philosophy and the Debate on Nuclear Weapons Systems and Policies."

This double issue of the journal has fourteen articles by philosophers from continental and analytic traditions on the nuclear themes of: (1) the extinction thesis, (2) power and domination, (3) nuclear weapons, and (4) the morality of deterrence and the quest for peace. The last section includes articles from the absolutist, just war, utilitarian, and pacifist perspectives. All contributors are critical of current nuclear weapons systems and policies, supporting positions from a freeze through unilateral nuclear disarmament.

Goodwin, Geoffrey, ed. *Ethics and Nuclear Deterrence.* New York: St. Martin's, 1982.

Eight essays, by British writers commissioned by the Council on Christian Approaches to Defense and Disarmament, use theological and ethical approaches to assess nuclear deterrence and the unilateralist and multilateralist routes to arms control and disarmament. A cross-section of Christian viewpoints, from deterrence as morally defensible to nuclear weapons themselves as "an offense to God" that must be eliminated even if by unilateral nuclear disarmament.

Green, Philip. *Deadly Logic: The Theory of Nuclear Deterrence.* Columbus, Ohio: Ohio State Univ. Pr., 1966.

Thorough-going critique and rejection of nuclear deterrence, especially as it has been developed theoretically in systems analysis and game theory. Strong ethical position permeates the analysis.

Hardin, Russell *et al.*, eds. *Ethics: An International Journal of Social, Political, and Legal Philosophy* 95 (April 1985): "Symposium on Ethics and Nuclear Deterrence."

The eighteen articles in this special issue arose from a three-day conference that brought together philosophers and strategists. Generally speaking, the philosophers represented here are from the analytic tradition (no continentalists or even just war theorists are represented) and the strategists are from the realist tradition. The arguments by strategists cover the main schools and those by philosophers can be divided between deontological and utilitarian approaches. A useful collection. Hardin and Mearsheimer do a good job in the introduction showing unifying themes and the diversity within and between the groups. Positions advocated range from more counterforce and strategic defense systems to nuclear disarmament.

Herzog, Arthur. *The War-Peace Establishment*. New York: Harper, 1963.

Herzog sought out and often interviewed a wide range of leading theoreticians on war and peace in the United States. He classifies these theoreticians into the deterrers, the experimentalists, and the peace movement. Whereas peace research falls under the second classification, the last is divided into survivalists and radicals (each with further divisions). Ends with a useful chart dealing with their respective views on power, probabilities of various events, and amount of time for finding solutions.

Hollenbach, David. *Nuclear Ethics: A Christian Moral Argument*. New York: Paulist Pr., 1983.

Author argues on theoretical and ethical grounds that for the pursuit of its ministry of justice and peace the Church must allow both pacifist and just war approaches to the morality of war. Also argues on ethical and prudential grounds that at present no use of nuclear weapons is justifiable and that concrete policies for deterrence must be evaluated individually in terms of whether they contribute to war prevention and disarmament.

Jaspers, Karl. *The Future of Mankind*. Translated by E. B. Ashton. Chicago: Univ. of Chicago Pr., 1961.

Originally published in German in 1958 as *Die Atombombe und die Zukunft des Menschen* (*The Atom Bomb and the Future of Man*), Jaspers's book, which elaborated on broadcasts he made in the fall of 1956, won the peace prize of the German book trade and also received sharp attacks. While Jaspers does present the disjunction of destruction or fundamental change, he does sanction some conditions under which destruction should be chosen. He opposes world government as the solution and, more like Kant, sees peace as more properly arising from a confederation. He sees pure coexistence as unlikely since it would require radical isolation. Given what he sees as the conflict between Soviet totalitarianism and Western freedom, he rejects survival if it would result in a totalitarianism in which people lose their humanity.

Myrdal, Alva. *The Game of Disarmament: How the United States and Russia Run the Arms Race*. Manchester: Manchester Univ. Pr., 1977.

In-depth analysis and critique of arms control and disarmament negotiations. Tries to develop an international point of view to identify and assess ''subterfuges and half-truths'' of the superpowers.

O'Brien, William V. *War and/or Survival*. Garden City, N.Y.: Doubleday, 1969.

Goes further than even Paul Ramsey in seeing, from a moral and theological perspective, a role for nuclear weapons. Though an argument for realism and against idealism, also in opposition to the Pax-American approach of Edward Teller, Robert Strauss-Hupe, and Curtis LeMay.

Ramsey, Paul. *War and the Christian Conscience: How Shall Modern War Be Conducted Justly?* Durham, N.C.: Duke Univ. Pr., 1961.

In this book by one of the leading theologians concerned with the just war tradition, Ramsey provides several chapters on the history of this tradition and on how it applies to the nuclear age. He has a chapter on nuclear testing and another which uses Kant to sketch a "Critique of Technical Reason" in which Ramsey shows the sterility of strategic studies in that logically they can lead to such antinomies as "deterrence is feasible" and "deterrence is not feasible" or "thermonuclear warfare is not feasible" and "thermonuclear warfare is feasible."

Russell, Bertrand. *Common Sense and Nuclear Warfare*. London: Allen & Unwin, 1959.

Considers various "isms" to be irrelevant to the danger of nuclear war and approaches the pursuit of peace at the highest level of common sense. Argues for not only the elimination of nuclear weapons but also the end to war, since during conflicts belligerents could remanufacture nuclear weapons. Stresses need for reduction of East-West tension so that any arms reductions will be significant (disarmament is "a palliative rather than a solution" [p. 47]). Also argues for an international authority. In one of the appendices, Russell notes how Soviet development of nuclear weapons led to his abandonment of his view at the time of the Baruch Proposal that the United States should pressure the Soviet Union to comply and, if necessary, threaten war in order to internationalize atomic weapons. This earlier view is expressed in his essay "The Future of Mankind" in *Unpopular Essays* (see Russell, 1950).

Schweitzer, Albert. *Peace or Atomic War?* New York: Holt, 1958.

The texts of three broadcasts by Schweitzer from Oslo, Norway on April 28, 29, and 30, 1958, on "The Renunciation of Nuclear Tests," "The Danger of an Atomic War," and "Negotiations at the Highest Level." Schweitzer cites scientific data on consequences of nuclear tests and war and calls for an end to testing and for nuclear disarmament.

Sterba, James P., ed. *The Ethics of War and Nuclear Deterrence*. Belmont, Calif.: Wadsworth, 1985.

Nineteen articles, reprints, and excerpts on general ethical theory, background information on effects of nuclear war and deterrence strategies, moral assessment of nuclear war and deterrence, and freeze vs. military buildup. Contributions by philosophers, scientists, politicians, and popularizers.

Wasserstrom, Richard A., ed. *War and Morality*. Belmont, Calif.: Wadsworth, 1970.

Includes several classic essays relevant to moral assessment of nuclear war, including William James's "The Moral Equivalent of War" and John C. Ford's "The Morality of Obliteration Bombing" (1944), as well as several essays that address nuclear weapons, such as Elizabeth Anscombe's "War and Murder" and Richard Wasserstrom's "On the Morality of War."

Allen, Verich S. and William V. O'Brien. *Christian Ethics and Nuclear Warfare*. Washington: Institute of World Policy, Georgetown Univ., 1963.

Aronson, Ronald. *The Dialectic of Disaster: A Preface to Hope.* London: Verso, 1983.

Batchelder, Robert C. *The Irreversible Decision: 1939–1950.* Boston: Houghton, 1961.

Benn, S. I. "Deterrence or Appeasement: Or, on Trying to Be Rational about Nuclear War." *Journal of Applied Philosophy* 1 (March 1984): 5–20.

Bennett, John C., ed. *Nuclear Weapons and the Conflict of Conscience.* New York: Scribner, 1962.

Boulding, Kenneth E. *Conflict and Defense: A General Theory.* New York: Harper, 1963.

British Council of Churches. *Christians and Atomic War: A Discussion of the Moral Aspects of Defense and Disarmament in the Nuclear Age.* London: British Council of Churches, 1959.

_____. *The Era of Atomic Power: Report of a Commission Appointed by the British Council of Churches.* London: SCM Pr., 1946.

Calhoun Commission. *Atomic Warfare and the Christian Faith: Report of the Commission on the Relation of the Church to the War in the Light of the Christian Faith Appointed by the Federal Council of the Churches of Christ in America.* New York: Federal Council of Churches, 1946.

_____. *The Christian Conscience and Weapons of Mass Destruction: Report of a Special Commission Appointed by the Federal Council of the Churches of Christ in America.* New York: Department of International Justice and Goodwill, Federal Council of Churches, December 1950.

Churchill, Robert P. "Nuclear Arms as a Philosophical and Moral Issue," *Annals, AAPSS* 469 (September 1983): 46–57.

Clancy, William, ed. *The Moral Dilemma of Nuclear Weapons.* New York: Council on Religion and International Affairs, 1961.

Dewey, John. "Dualism and the Split Atom: Science and Morals in the Atomic Age," *The New Leader* 28 (November 1945): 1, 4.

Dimler, G. Richard, ed. *Thought* 59 (March 1984); "The Morality of Nuclear Deterrence."

Dombrowski, Daniel. "Gandhi, Sainthood, and Nuclear Weapons," *Philosophy East and West* 33 (1983): 401–6.

Driver, Christopher. *The Disarmers.* London: Hodder & Stoughton, n.d.

Falk, Richard A. *Law, Morality, and War in the Contemporary World.* New York: Praeger, 1963.

_____ and Richard I. Barnet, eds. *Security in Disarmament.* Princeton, N.J.: Princeton Univ. Pr., 1965.

Gauthier, David. "Deterrence, Maximization, and Rationality," *Ethics* 94 (1984): 474–95.

Gay, William C. "Myths about Nuclear War: Misconceptions in Public Beliefs and Governmental Plans." *Philosophy and Social Criticism* 9 (Summer 1982): 115–44.

Gessert, Robert A. and J. Bryan Hekir. *The New Nuclear Debate.* New York: Council on Religion and International Affairs, 1976.

Geyer, Alan. *The Idea of Disarmament! Rethinking the Unthinkable.* Elgin, Ill.: Brethren Pr., 1982.

Ginsberg, Robert. *The Critique of War: Contemporary Philosophical Exploration.* Chicago: Regnery, 1969.

Gollancz, Victor. *The Devil's Repertoire of Nuclear Bombing and the Life of Man.* Garden City, N.Y.: Doubleday, 1959.

Goodin, Robert E. "Disarming Nuclear Apologists." *Inquiry* 28 (1985): 153–76.

Govier, Trudy. "Nuclear Illusion and Individual Obligations." *Canadian Journal of Philosophy* 13 (December 1983): 471–92.

Hardin, Russell. "Unilateral versus Mutual Disarmament." *Philosophy and Public Affairs* 12 (Summer 1983): 236–54.

Heim, Michael. "Reason as Response to Nuclear Terror." *Philosophy Today* 28 (Winter 1984): 300–7.

Hook, Sidney. *The Fail-Safe Fallacy.* New York: Stein & Day, 1963.

Johnson, James Turner. *Can Modern War Be Just?* New Haven, Conn.: Yale Univ. Pr., 1984.

_____. *Just War Tradition and the Restraint of War: A Moral and Historical Inquiry.* Princeton, N.J.: Princeton Univ. Pr., 1981.

Kaplan, Morton A., ed. *Strategic Thinking and Its Moral Implications.* Chicago: Univ. of Chicago Center for Policy Studies, 1973.

Kavka, Gregory S. "Deterrence, Utility and Rational Choice." *Theory and Decision* 12 (March 1980): 41–60.

_____. "Some Paradoxes of Deterrence," *The Journal of Philosophy* 75 (June 1978): 285–302.

Kenny, Anthony. *The Logic of Deterrence.* London: Firethorn Pr., 1985.

Keys, Donald, ed. *God and the H-Bomb.* New York: Bellmeadows, 1961.

Lackey, Douglas P. "Ethics and Nuclear Deterrence." in *Moral Problems,* edited by James Rachels (New York: Harper, 1975), pp. 332–45.

_____. "Missiles and Morals: A Utilitarian Look at Nuclear Deterrence." *Philosophical and Public Affairs* 11 (Summer 1982): 189–231.

_____. *Moral Principles and Nuclear Weapons.* Totowa, N.J.: Rowman & Allenheld, 1984.

Lawler, Justus George. *Nuclear War: The Ethic, the Rhetoric, the Reality: A Catholic Assessment.* Westminister, Md.: Newman Pr., 1965.

Levine, Robert A. *The Arms Debate.* Cambridge, Mass.: Harvard Univ. Pr., 1963.

MacLean, Douglas, ed. *The Security Gamble: Deterrence Dilemmas in the Nuclear Age.* Totowa, N.J.: Rowman & Allenheld, 1984.

McMurrin, Sterling M., ed. *Values at War: Selected Tanner Lectures on the Nuclear Crisis.* Salt Lake City, Utah: Univ. of Utah Pr., 1983.

Merton, Thomas, ed. *Breakthrough to Peace: Twelve Views on the Threat of Thermonuclear Extermination.* New York: New Directions, 1962.

Methodist Church. *The Christian Faith and War in the Nuclear Age.* New York: Abingdon, 1963.

Milford, T. R. *The Valley of Decision: The Christian Dilemma in the Nuclear Age.* London: British Council of Churches, 1961.

Murray, Thomas. *Nuclear Policy For War and Peace.* Cleveland, Ohio: World, 1960.

Nagle, William J. *Morality and Modern Warfare: The State of the Question.* Baltimore: Helicon Pr., 1960.

National Conference of Catholic Bishops. "The Challenge of Peace: God's Promise and Our Response," *Origins* 13 (May 19, 1983): 1–32.

Newman, James. *The Role of Folly.* London: Allen & Unwin, 1962.

O'Brien, William V. "Just War in a Nuclear Context," *Theological Studies* 44 (1983): 191–220.

_____. *Nuclear War, Deterrence and Morality.* Westminster, Md.: Newman Pr., 1967.

Palter, Robert M. "The Ethics of Extermination," *Ethics* 74 (April 1964): 208–18.

Paskins, Barrie and Michael Dockrill. *The Ethics of War.* Minneapolis: Univ. of Minnesota Pr., 1979.

Paul, Ellen Frankel, ed. *Social Philosophy and Policy* 3 (Autumn, 1985); "Nuclear Rights/Nuclear Wrongs."

Potter, Ralph B. *War and Moral Discourse.* Richmond, Va.: John Knox Pr., 1969.

Ramsey, Paul. *The Just War: Force and Political Responsibility.* New York: Scribner, 1968.

_____. *The Limits of Nuclear War: Thinking about the Do-Able and the Un-Do-Able.* New York: Council on Religion and International Affairs, 1963.

Rapaport, Anatol. *Fights, Games and Debates.* Ann Arbor, Mich.: Michigan Univ. Pr., 1960.

_____. *Strategy and Conscience.* New York: Harper, 1964.

Roszak, Theodore. "A Just War Analysis of Two Types of Deterrence." *Ethics* (January 1963): 100–9.

Russell, Bertrand. *Has Man a Future?* New York: Simon & Schuster, 1962.

_____. *Unpopular Essays.* New York: Simon & Schuster, 1950.

Santoni, Ronald E. " 'Just War' and Nuclear Reality," *Philosophy Today* (1985): 175–90.

_____. "Nuclear Insanity and Multiple Genocide." In *Towards Understanding, Intervention and Prevention of Genocide,* edited by Israel Charney (Boulder, Colo.: Westview, 1984).

Schell, Jonathan. *The Abolition.* New York: Knopf, 1984.

_____. *The Fate of the Earth.* New York: Knopf, 1982.

Schilpp, Paul Arthur. "A Challenge to Philosophers in the Atomic Age." *Philosophy* 24 (January 1949): 56–68.

Schorstein, Joseph. "The Metaphysics of the Atom Bomb." *Philosophical Journal* 5 (1962): 33–46.

Shaw, William H. "Nuclear Deterrence and Deontology." *Ethics* 94 (January 1984): 248–60.

Shibata, Shingo. "The Right to Life vs. Nuclear Weapons." *Journal of Social Philosophy* 8 (September 1977): 9–14.

Sider, Ronald J. and Richard K. Taylor. *Nuclear Holocaust and Christian Hope: A Book for Christian Peacemakers.* Downers Grove, Ill.: Intervarsity Pr., 1982.

Singer, J. David. *Deterrence, Arms Control and Disarmament: Towards a Synthesis in National Security Policy.* Columbus, Ohio: Ohio State Univ. Pr., 1960.

Smithka, Paula J. "Heidegger and Nuclear Weapons." *Concerned Philosophers for Peace Newsletter* 7 (April 1987): 8–10.

Somerville, John. "The Contemporary Significance of the American Declaration of Independence." *Philosophy and Phenomenological Research* 38 (June 1978): 489–504.

————. "Nuclear Omnicide: Moral Imperatives for Human Survival." *New World Review* (1983): 20–21.

————. *The Peace Revolution: Ethos and Social Process.* Westport, Conn.: Greenwood, 1975.

————, ed. *Soviet Marxism and Nuclear War: An International Debate.* Westport, Conn.: Greenwood, 1981.

————. "Soviet Marxism in Today's Perspective." *Journal of Social Philosophy* 7 (September 1976): 7–11.

Stein, Walter, ed. *Morals and Missiles: Catholic Essays on the Problem of War Today.* London: James Clarke, 1959.

————. *Nuclear Weapons: A Catholic Response.* New York: Sheed & Ward, 1961.

Stevenson, Leslie. "Is Nuclear Deterrence Ethical?" *Philosophy* 61 (1986): 193–214.

Teilhard de Chardin, Pierre. *The Future of Man.* New York: 1964.

Tucker, Robert Warren. *The Just War: A Study in Contemporary American Doctrine.* Baltimore: Johns Hopkins Univ. Pr., 1960.

Vanderhaar, Gerard A. *Christians and Nonviolence in the Nuclear Age: Scripture, the Arms Race, and You.* Mystic, Conn.: Twenty-Third Publications, 1982.

Wallis, Jim, ed. *Peace-Makers: Christian Voices from the New Abolitionist Movement.* San Francisco: Harper, 1983.

————. *Waging Peace: A Handbook for the Struggle against Nuclear Arms.* New York: Harper, 1982.

Walzer, Michael. *Just and Unjust Wars.* New York: Basic Books, 1977.

Wells, David. "How Much Can 'The Just War' Justify?" *Journal of Philosophy* 66 (December 1969): 819–29.

————. *War Crimes and Laws of War.* Washington, D.C.: University Pr. of America, 1984.

Wolff, Robert Paul. "Maximization of Expected Utility as a Criterion of Rationality in Military Strategy and Foreign Policy," *Social Theory and Practice* 1 (Spring 1970): 99–111.

Zahn, Gordon C. *War, Conscience and Dissent.* New York: Hawthorn, 1967.

Zimmerman, Michael. "Humanism, Ontology, and the Nuclear Arms Race." In *Research in Philosophy and Technology,* VI, edited by Paul T. Durban and Carl Mitcham. Greenwich, Conn.: Jai Pr., 1983.

CHAPTER 8

Alternative Futures:
War, Weapons,
and National Security

To understand future possibilities, one needs to develop a deeper sense of where we have been. We have focused on the existence of nuclear weapons and the policies guiding their use; the consequences of the nuclear arms race; and the responses voiced by policy makers, strategists and critics. Before moving toward a discussion of alternative futures, we briefly re-capitulate some of the major ideas/patterns discussed in this text (more detailed discussion is found in the relevant chapters). Eight major dimensions have been discussed.

First, war in the twentieth century has been distinctive in character (see Chapter 1). Total war has become possible—unlimited war in which all resources of a country are employed against the total national life of an enemy. This possibility has led developed countries to institutionalize war preparation. It also has led to important alterations in the rules of war. Today the distinction between combatant and noncombatant and the understanding of limited military objectives and limited force have been altered. The introduction of nuclear weapons has intensified these changes: the superpowers continue to build weapons of mass destruction to use against civilian and military targets and to threaten massive retaliation.

Second, nuclear weapons have evolved from rudimentary atomic bombs capable of causing 150,000 deaths to highly accurate thermonuclear weapons, each capable of killing millions (see Chapters 2 and 5). These weapons, incorporated in TRIAD systems, can be delivered by bomber, missile, or submarine. Nuclear weapons have become incorporated within traditional military units. Qualitative improvements have outstripped quantitative developments: with improved missile accuracy and the development of MIRVs, use of nuclear weapons against military targets becomes an operational possibility. The United States has announced plans to significantly increase the number and quality of its nuclear arsenal, which has intensified the debate that characterizes most changes in existing arsenals. A minimum of five countries now possess nuclear weapons, with another five to fifteen countries plausibly capable of becoming nuclear club members. Horizontal and vertical proliferation has reemerged as significant issues.

Third, policies (official statements and understandings about the role of nuclear weapons in national security) have evolved from containment through threatened use of weapons, to massive retaliation and assured destruction, to varying statements regarding flexible response (see Chapter 3). A policy of first use of nuclear weapons in Europe to compensate for inferior troop strength and the recently announced interest in developing a defensive system continue to draw strong criticism.

Fourth, nuclear policies are related to the evolution of nuclear planning (see Chapter 4). This planning, predicated in varying degrees on probability and formal strategy, involves government scenarios (plans for use and targets). The evidence indicates that initial U.S. nuclear planning was oriented toward first-strike use within a total war orientation. Since 1960, the United States has operated under a SIOP (single integrated operational plan), replacing redundant and competing nuclear plans of the armed services. This SIOP, which has undergone numerous revisions, has vacillated in stressing countervalue (people) or counterforce (military) targets, strategic or theater nuclear options, and controlling or winning a nuclear war. The United States currently designates almost 40,000 Soviet military and civilian targets.

Five, our understanding of the consequences of nuclear detonations has increased in some areas but remains speculative in others (see Chapter 5). We generally understand that a nuclear exchange would result in destruction unknown in previous wars; the extent of the destruction remains a matter of debate. We understand better the long-term medical, biological, and environmental consequences of a ground zero explosion. Certain consequences remain matters of speculation: electromagnetic pulse, ozone depletion, nuclear winter, and so on. We know that survival is likely, yet the level and the quality of survival continues to be debated.

Six, the development of nuclear arsenals has occurred as part of a global pattern of militarization and increased reliance on sophisticated weaponry (see Chapter 6). While the Soviet Union and United States have relied heavily on systems of nuclear weapons, the rest of the world has become more dependent on increasingly sophisticated conventional armaments. The conventional and nuclear arms races have facilitated global inflation, dramatically influenced peacetime military budgets, drained scientific resources, and redirected economic activity.

Seventh, a large number of nuclear specialists or experts have emerged (see Chapter 7). Early specialists saw the significance of nuclear weapons and identified their major problems, such as proliferation and the strategy of deterrence; yet these early writers had little initial effect on government policy. Later these writers would have a direct and lasting effect on the development of strategic studies within academia and on policy through their writings on deterrence theory, a subset of strategic theory. While deterrence theory, in varying forms, remains a dominant idea today, it has been under attack by revisionists (who claim deterrence is not a strategic plus, as it is based on a balance of terror rather than a policy for winning) and religious and philosophical critics (who argue that deterrence is either a flawed or simply immoral concept). The revisionist line has been generally accepted by the Reagan administration.

Eighth, the immediate past has been characterized by stalled arms control talks, an emphasis on new weapons development, and a move away from the theory of deterrence toward a Strategic Defense Initiative (SDI)—an orientation toward defense, rather than mutual terror (see this chapter and Chapter 7). Concern and criticism directed towards these and other policy directions have led to intensified debate over such issues as cruise missiles, Euromissiles, MX, SDI, nuclear freeze, and so on. Almost all of these issues may be traced to the competing models of warfighting and war-deterring and to the role nuclear weapons play in either model. The strategic community tends to accept the existence of nuclear weapons but differs on their function. Recent concern over nuclear policy has however brought up the question, are there security models not dependent on nuclear weapons?

In addressing these eight topics we have shown the forty-year history of the nuclear debate. During the past few years, the entire dependency on nuclear weapons has come under renewed attack, and it is this recent attack and future options that we consider in this chapter. The recent attack came from those backing the idea of nuclear fighting. Throughout much of the 1970s, there was a concentrated attack on the policy of deterrence. A strategy of winning was needed. Additionally, as has occurred cyclically, there was a political argument that our strategic forces were vulnerable because of Soviet spending and modernization and other factors. Beginning in the late 1970s, there were increased demands for policy change. The attack on deterrence worked: countervailing plans were developed, funds expended for new weapons systems, and a move made toward strategic defense.

Reaction to the planned and announced proposals for modernizing and expanding the nuclear arsenals has been strong and pervasive. Experts in strategic studies, academicians, and policymakers alike, have heavily criticized the apparent rejection of MAD and adoption of a nuclear fighting doctrine. Groups of concerned citizens—including Union of Concerned Scientists, Ground Zero, Federation of American Scientists, European Nuclear Disarmament, United Campuses to Prevent Nuclear War, SANE, and Physicians for Social Responsibility (PSR)—formed or rapidly increased their membership to lobby and educate the public about the dangers of the new policies. Individuals, including Helen Caldicott and Jack Geiger (both of PSR) and Carl Sagan, became important speakers at rallies and protest demonstrations. Jonathan Schell's best-seller *The Fate of Earth* warned of an apocalypse and called for an end to national sovereignty as the only alternative to nuclear war. Hundreds of thousands of citizens participated in rallies and protests, demanding a change in national military and foreign policy. Movies with topics such as accidental nuclear war and the futility of postattack survival were seen by tens of millions. Organized protest emerged in Western Europe in response to American plans to deploy nuclear missiles. Religious groups issued statements condemning the arms race and demanding disarmament. High schools and colleges began offering courses on nuclear war. In short, concern over nuclear weapons and the threat of nuclear war became *the* political issue of the early 1980s. It is difficult to overstate the rapidity with which nuclear issues entered the public arena. Debates have focused on American policy, weapons, arms control, and the possible consequences of nuclear war.

In the area of general policy, the debates have centered on three issues. First, there has been widespread disagreement among nuclear specialists and the general public over the deployment of cruise and Pershing missiles in Europe. For the first time in twenty years, American missiles were stationed in Great Britain. NATO policy with regard to use of nuclear weapons in a regional conflict became a major issue in France, Great Britain, Italy, and West Germany. Varying attempts were made to block the deployment of American missiles. Second, concern was expressed about adoption of PD-59: were we planning for limited or protracted nuclear war in Europe? Third, and most recently, debate has centered on the proposed Strategic Defense Initiative, a seeming repudiation of the prohibition of nuclear weapons in space, and a new catalyst in the arms race.

Debates were also waged over decisions to greatly expand and modernize the nuclear arsenal. The development and deployment of the Trident II submarine, the B-1 bomber, and the cruise missile, as well as the planned production of the MX, the stealth bomber, and space weapons increased public concern over why more weapons were needed and whether they were first-strike weapons. The cost associated with production and development of these weapons in a time of budget deficits and scarce tax dollars also became an issue. The initial rationale for these new weapons was to close a window of vulnerability to the strategic TRIAD; more recently, it has been argued that many of these weapons represent "bargaining chips" in arms negotiation with the Soviet Union.

The war fighters' attack on defense policy of the late 1970s as weak and misdirected helped block consideration of the SALT II treaty. The current administration, while critical of SALT II, until recently agreed to abide by its weapons limitations. At the same time, it has argued instead for the negotiation of START proposals. For most of the past five years, there have been little if any arms control negotiations. Arms control has traditionally been an integral element of the war deterrers' position; its advocates have been among the most vocal in demanding progress in arms control. Alternatives to the current perspective are most clearly seen in the proposals for a nuclear freeze, a no first use policy, and a comprehensive test ban treaty.

Debates also emerged around the possibility of accidental nuclear war. This concern began with the accident at Three Mile Island and was intensified by the Chernobyl disaster. It is also influenced by nuclear bomb accidents, the possibility of nuclear terrorism, and the realization that with the deployment of missiles in Europe, the response time to potential nuclear attacks had been reduced from thirty to as little as six minutes.

Finally, a debate focused on the consequences of the nuclear arms race and nuclear war. On the one hand, funding for the $1 trillion military buildup came at a time when social programs in the United States were being cut. Public opinion polls supported the strengthening of the military, but differed strongly over which kinds of modernization were most needed. As the most rapidly increasing proportion of military spending, nuclear programs came under attack as unnecessary and potentially destabilizing in their effect on Soviet-American relations. While speculative, fears of ozone depletion and nuclear winter intensified public anxiety and criticism. Existing or proposed plans involving civil defense were ridiculed as unworkable and resulting in a false sense of security about survival in a nuclear war.

The debate over nuclear weapons and nuclear war may be seen as a reflection of three visions of politics and the role of the military. We may phrase this in a series of distinctions: (1) views of national sovereignty and global security (idealist vs. realist); (2) views toward the military role in maintaining sovereignty (conventionalists vs. nonconventionalists); (3) views toward current debates (particularly policies, weapons, and arms control); (4) visions of the future (including the arms race, the possibility of war and its consequences, and the future global role of the United States).

These distinctions lead to three groups of approaches. In the remaining three sections we distinguish between those arguing for a nuclear future, those arguing for a conventional (nonnuclear) future, and those arguing for a nonmilitary future (without weapons or war). These approaches are not exclusive, as some writers could be categorized in more than one approach. But the intent here is to shed light on the current debates and extrapolate these positions to future visions. A summary of the approaches is found in table 10.

Table 10.

Four Perspectives on Nuclear Weapons and War

Approaches to War and Weapons	Sovereignty and Global Security	Role of the Military in Sovereignty	Current Debates	Vision of the Future
Nuclear deterrers (Brodie, McNamara, Catholic bishops)	Global security possible through strong sovereignty and alliances, emphasis on bipolar *parity* and *stability*.	Support strong military in pursuit of national security. Have sufficient nuclear forces, need buildup of conventional forces.	Support arms control and efforts to maintain parity and stability. Oppose new weapons as destabilizing.	Continued balance of terror; continue but control arms race. Nuclear war possible.
Nuclear fighters (Kahn, Gray, Reagan)	To maintain sovereignty, weapon *superiority* is crucial; must deal from strength, not parity.	Support strong military in pursuit of national security. Need new weapons to remain dominant (e.g., MX, SDI).	Strategy of prevailing in protracted conflict. Deterrence and arms control harmful. Advocate BMD.	Perpetual confrontation with USSR. Continued arms race. Political survival more important than effects of nuclear war.
Conventional deterrers (United Nations, CDI Physicians for Social Responsibility)	Maintain sovereignty and global security without nuclear weapons; seek multipolar world.	Support strong military in pursuit of security, but seek regional and international political alternatives.	Reject nuclear arms race. Support arms control, emphasis on economic development.	Multipolar world. Nuclear disarmament. Strengthened international organizations. Conventional warlike, but controlled.
Pacifists (Woito, Clark and Sohn, Sharp)	Includes those emphasizing nonviolence and civil disobedience, world government, and civilian defense.	Strong military not necessary in pursuit of security; emphasis on alternative mechanisms.	Reject weapons and war. Support arms control, civilian defense, peacemaking.	World without war. Alternative means of conflict resolution. If war, resist.

Living with Nuclear Weapons

The prevailing perspective on nuclear weapons is encompassed by those arguing that we must learn to live with nuclear weapons. The scientific knowledge exists for their production; no compelling arrangement exists for their elimination (contrary to international agreements), and they may be seen as having been successful in avoiding war between the superpowers during the past forty years. Within this approach, however, are the two significant subgroups discussed in Chapter 7—those that see the function of nuclear weapons as furthering the deterrence of war, and those that see nuclear weapons as important tools in furthering national policy (superiority). Both groups take a realist view of politics (that military force is necessary for national security); more specifically, they see a continuing role for nuclear weapons as part of military and political policy.

Nuclear Deterrers

Nuclear deterrers favor a strong military posture in pursuit of national security. While they support a continued role for nuclear forces, they are uneasy with certain quantitative and qualitative directions in the nuclear arms race and wish to see them curbed. For nuclear deterrers, the number of nuclear weapons stockpiled is generally sufficient. While the nuclear deterrers favor maintaining these stockpiles (ensuring that the weapons and their delivery vehicles are operative and secure), they do not perceive increases in the numbers of weapons as enhancing security. Likewise, they do not perceive further refinements of missile accuracy as enhancing security; rather, increased accuracy is viewed as making a first strike more likely. At the same time, they do support increasing the quantity and quality of conventional forces, particularly in Europe. These positions follow from their version of political realism. Their view of military security accepts the premise that stability flows from a balance of forces. As a result, they accept nuclear parity with the Soviets as desirable. Such a military balance, they contend, best holds in check ideological differences between the superpowers. Both the United States and the Soviet Union are deterred from using their nuclear forces to the extent to which each side perceives itself as lacking the kind of superiority necessary for successful war-fighting.

They view the current debates over policy, weapons, and arms control from the perspective of stability—that is, do changes weaken or strengthen stability? Is a new system destabilizing to superpower relations? Interestingly, in current debates, it is not always easy to distinguish nuclear deterrers from those who are altogether opposed to nuclear weapons. For example, nuclear deterrers oppose the MX missile, the Pershing missile, and the current Strategic Defense Initiative (the so-called Star Wars program) as being potential first-strike weapons (and hence destabilizing). They approve of arms control as a means of maintaining parity or constraining future weapon developments. They support ratification of SALT II, and many support such proposals as no first use, but generally agree that a nuclear freeze is potentially destabilizing (see Chapter 7).

Their vision of the future is a continuation of the balance of terror; they see the nuclear arms race as continuing but in a controlled manner. They support efforts to control the spread of nuclear weapons to other nations, arguing that the current balance is manageable. They think that through serious diplomacy with the Soviets and some luck, nuclear war can be avoided. However, nuclear deterrers believe that if negotiations break down or war-fighting policies and capabilities are adopted, then our future is more likely to result in a nuclear war that will destroy civilization as we know it.

The viewpoint of the nuclear deterrers is discussed at length in the first three sections of Chapter 7, and the interested reader can consult that portion of the text and its bibliography for many references. Only a few further authors and issues are cited here. Brodie offers a defense of nuclear deterrence; he argues that nuclear weapons deter not only nuclear but also conventional wars between the superpowers. McNamara takes the limit of nuclear weapons about as far as it can go. He claims *"Nuclear weapons serve no military purpose whatsoever. They are totally useless—except only to deter one's opponent from using them."*[1] Somewhat between these two positions is that of Jervis. In his critique of countervailing strategy, he warns against thinking nuclear weapons can be used like conventional ones and of the dangers of "coupling" NATO and U.S. strategic forces, and ends defending minimum deterrence. Finally, but only as an interim position, the pastoral letter by the National Conference of Catholic Bishops accepts a version of the nuclear deterrer position.

Nuclear Fighters

Nuclear fighters share with the nuclear deterrers the view that a strong military posture is necessary. In their version of political realism, however, a nation must aim for superiority, not parity. As a result, they favor quantitative and qualitative increases in nuclear and conventional forces. They believe that the current parity of forces with the Soviet Union forces the United States into a position of self-deterrence—the United States cannot use its forces without threatening a nuclear confrontation. Nuclear fighters accept the role of superpower in the world and emphasize the necessity to remain dominant. They see the policy of deterrence as a hindrance to the pursuit of international political objectives. Insofar as the United States is militarily superior to other nations, its influence can be guaranteed. Specifically, nuclear fighters want the Soviet military to be inferior to the U.S. military to increase the possibilities that either the Soviets will make concessions to the United States rather than go to war, or the Soviets will suffer defeat if they do go to war. Therefore, a strategy of winning is necessary.

Emphasizing strength, the nuclear fighters support modernization of existing weapons and the development of new systems. They argue that the United States must be willing to stay ahead in the arms race. They vigorously support such weapons as the neutron bomb, the cruise missile, the stealth bomber, and the MX, and strongly support the idea of the

---

1. Robert McNamara, "The Military Role of Nuclear Weapons: Perceptions and Misperceptions," *Foreign Affairs* 62 (1983): 79.

Strategic Defense Initiative. They oppose most forms of arms control, unless the United States is the clear winner in the negotiations. They view the antinuclear movement as dangerous, the deterrers as appeasers, and the Soviets as untrustworthy, ideologically abhorrent, and militarily provocative. Nuclear fighters have provided an influential critique of deterrence and its relation to foreign policy objectives. They tend to emphasize the importance of NATO commitments and forward basing (putting troops and weapons close to the enemy). Their views, once considered too bellicose, are now increasingly acceptable.

Their vision of the future sees the United States and the Soviet Union locked into a perpetual confrontation; any lapse in vigilance would be fatal. They believe that political survival is more important than the possible consequences of a nuclear confrontation. They see no prospect that the arms race will end and argue that the Soviet Union gains from efforts at control. Nothing can guarantee that war will not erupt; in the event that it does, the nuclear fighters believe that the United States must be prepared to fight and win. Hence, they support planning for protracted nuclear fighting and the idea of civil defense. In other words, they want U.S. dominance with or without war: in case of war, they believe counterforce weapons, ballistic missile defense systems, and civil defense will enhance survival for the United States. They are prepared to accept the destruction of nuclear war rather than capitulate to Soviet communism.

The viewpoint of the nuclear fighters is discussed at length in the first three sections of Chapter 7, and the interested reader can consult that portion of the text and its bibliography. Only a few further authors and issues are cited here. The extent of their policy influence is reflected in the U.S. Department of Defense Annual Reports by Brown (1981) and Weinberger (1983). The two reports advocate escalation dominance, which entails a nuclear war-fighting capability. Gray continues to be one of the most outspoken advocates of this position. He recently claimed "the United States needs both to be able to deny victory to the Soviet Union and—no less important—to avoid defeat itself. These requirements add up to a requirement for a capability to win wars."[2]

Closely associated with this school is the call for space weapons: antisatellite and ballistic missile defense systems. It is argued that such a policy, if implemented, would obviate the need for more offensive weapons and constitute a more meaningful deterrent. The Strategic Defense Initiative is reflected in the works by Graham and Carter and Schwartz. For congressional assessment, see House Committee on Armed Services (1984). For a critical assessment, see Stein and Union of Concerned Scientists.

2. Colin Gray, "Warfighting for Deterrence," *Journal of Strategic Studies* 7 (1984): 11.

Living without Nuclear Weapons

The emphasis in this section is on conventionalists—those persons who consciously reject nuclear weapons or believe that conventional security systems are adequate. Included here are both conventional war deterrers and conventional war fighters. Persons taking such a perspective normally support a strong military posture and conventional arms trade. They define themselves as political realists, only they rely on nonnuclear forces.

Supporters of this position argue that conventional weapons are more likely than nuclear weapons to facilitate national objectives. Support for this position comes from representatives of both nonnuclear and nuclear nations. Particular support can be found in the signatories to the Nonproliferation Treaty, the United Nations, and antinuclear organizations such as SANE and Physicians for Social Responsibility. It should be noted that although some of these organizations ultimately favor conventional disarmament as well, they are included here insofar as they support conventional military forces, at least as an interim stage in the quest for national security.

Consistent with this approach is a rejection of the nuclear arms race and its attendant development of new weapon systems (MX, cruise missiles, Trident II, and so on). Conventionalists support efforts at nuclear arms control and push for actual arms reductions. They advocate an important role for international organizations in any negotiations, arguing that nuclear weapons represent global concerns. The Third World countries emphasize the correlation between the nuclear arms race and slowed development in the poorer states. Switzerland, a developed nation, offers a somewhat unique alternative perspective. It leads the world in per capita military expenditures, but it neither has nor wants nuclear weapons. It has an exclusively defensive-oriented system, saving its highest awards for military leaders who kept it out of war.

Conventionalists envision a future in which nuclear weapons have been controlled or eliminated, a vision allowing national sovereignty through the establishment of a multipolar rather than bipolar world. They envision a series of possible regional security systems rather than a strict dependence upon either a large standing military organization or inclusion under a nuclear umbrella. They recognize that regional conflicts may persist and limited wars still be waged, but they believe that nations not party to these wars will at the least be spared the kind of indirect consequence that might befall them in the case of wars fought with nuclear weapons.

The conventional deterrence and war-fighting perspective encompasses more than models for exclusively conventional military organization. Many writers on conventional weapons, policies, and strategies see them as part of the broader arrangement that includes the control of nuclear weapons, and they do not question whether conventional forces can and should be sufficient. Helpful studies on conventional deterrence can be found in Quester, Mearsheimer, Kaufmann, and Heilbrunn. Many of the articles in journals such as *Peace Research* and *Journal of Conflict Resolution* seek to examine the mechanisms necessary to move away from a nuclear world. One group that stands out in its clear advocacy of a strong military but strong criticism of nuclear weapons is the Center for Defense Information. For discussions of conventional defense in Europe, see Komer and Golder et al. Also relevant is the work of Heyer and the Alternative Defense Committee. The largest number of works are issued by the United Nations (see Appendix for examples of their publications).

Living without War

We designate as pacifists those who reject violence and war. Unlike the conventionalists or the nuclear deterrers or fighters, pacifists are political idealists. They believe agreements can be binding without the threat of force, and if an agreement is broken, they believe there are nonviolent ways to respond. Pacifists reject the realist premise that a strong military posture is necessary for national security. They stress instead various nonmilitary routes to enhanced security. For example, they typically emphasize both at home and abroad strong moral commitments to human rights and social justice and see the moral capital of such a posture, as well as the pursuit of economic well-being for all nations, as more conducive to national security than the accumulation of nuclear and conventional arms.

Pacifists view the current debates over policy, weapons, and arms control from the perspective of peacemaking—that is, they consider the extent to which policies threaten violence, weapons facilitate the execution of violence, and arms control proposals militate against the prospects for violence. As a result, they oppose each policy variation on proper military posture and each program for the development, production, and deployment of a new weapon of violence. For example, they vigorously oppose the neutron bomb, the MX, and the cruise missile. Nevertheless, as interim steps they are likely to support most arms control proposals as preferable to uncontrolled arms races. Pacifists typically support the nuclear freeze, no first use, a comprehensive test ban, and nuclear free zones on earth and in space.

The pacifist vision of the future is one in which the peoples of the earth have finally learned to live together in peace and without war. Pacifists see no international or personal adversaries as so evil that reconciliation and cooperation are precluded. They are not Manicheans—they do not believe good resides in any one person or nation and evil in any one person or nation. They believe all nations and persons have shortcomings, and can work together nonviolently to improve themselves and others. Ultimately, moral commitments to nonviolence are more important than any political value, including the maintenance of national sovereignty. While they believe the arms race can end and disarmament can occur, they are prepared to respond only nonviolently if war breaks out before their vision is realized.

Alternatives to war have been proposed about as long as critiques of war have been made. From Lao-Tzu to Martin Luther King, Jr., religious and social leaders around the world have pointed to nonviolent forms of conflict resolution. Advocates of a warless world admit that effective military deterrence, whether nuclear or conventional, entails the capacity for, and often the policy of, retaliation. Nevertheless, those who want to go beyond war want to go beyond the *ad baculum* (big stick) form of persuasion that military deterrence and force perpetuate. Mayer and Weinberg and Weinberg have useful historical anthologies on the rejection of war and violence. Woito provides one of the most helpful bibliographic sources on approaches to ending war.

The perspectives on how to live without war can be subsumed under three prevalent themes: (1) nonviolence and civil disobedience, (2) international law and world government, and (3) civilian defense. Although these themes are not mutually exclusive, they suggest the general topics around which most of the debates are clustered. In this regard, Beitz and Herman and Ginsberg have anthologies that contain articles on several of these approaches.

Nonviolence and Civil Disobedience

In the twentieth century, nonviolence and civil disobedience are most closely associated with Mahatma Gandhi and Martin Luther King, Jr. Their pacifism was hardly a capitulation before, or accommodation of, unjust social forces. Rather, they forged nonviolent methods for social change. Their nonviolence arose from courage, not fear, and was based on moral, not pragmatic, principles. Accepting but not inflicting suffering, they sought dialogue, turning to demonstrations and civil disobedience as a last resort. They held constantly to nonviolence as the only means. As a result, they varied their goals to ones that could be achieved by their means. When thwarted in these endeavors, they had constructive work (such as community projects), another component of their methods, as a way of positively redirecting frustration. Gandhi and King have both provided texts in which they detail these methods. Bedau has a good collection of essays on civil disobedience, and Bondurant and Horsburgh have written helpful introductions to Gandhian philosophy.

The approaches of Gandhi and King, of course, were not forged to end war per se, although advocates of *civilian defense* draw heavily on their works. (The extension of these methods to deal with war and occupation are discussed in the final subsection). Moreover, their method is often associated with its opposite. As Woito notes:

> The transition in the civil rights and peace movements from a commitment to principled nonviolence to mass movements in which violence was tolerated and sometimes encouraged, served to discredit and obscure the value of nonviolent theory.[3]

What Gandhi and King did show was that violence could be resisted and even countered by nonviolence. Hence, neither the use nor the threat of violence is necessary for social change.

International Law and World Government

Twentieth-century calls for international law and world government as means to end war are closely associated with the advent of nuclear weapons, although World War I also motivated many people to look in this direction as well for a solution to the problem of war (for example, League of Nations and World Federationists). One of the first calls in the nuclear age came from Cousins less than two weeks after the atomic bombings of Hiroshima and Nagasaki. More well-known, however, are the efforts of Clark and Sohn. Both works advocated increased reliance on the development of international organizations to resolve conflict.

Insofar as nation states remain intact, means for resolving their conflicts continue. One of the frequently cited means is Graduated Reciprocation in Tension-Reduction (GRIT) proposed by Osgood. The United States would seek to reverse the arms race by making unilateral initiatives toward tension reduction and disarmament as long as the Soviets responded in kind. A similar approach can be found in Fisher. The World Policy Institute (formerly the Institute for World Order) has undertaken a series of reports delineating models of nonviolent international organization; see the works of Falk and Mendlovitz, Falk, Mendlovitz, Johansen, and Beres and Targ; Wein provides a compilation of programs and courses oriented toward the goals of world order.

Civilian Defense

One of the more intriguing approaches to a world without war is civilian defense, sometimes termed nonviolent national defense or civilian resistance. Civilian defense entails prior pledge and subsequent performance by citizens of organized nonviolent resistance to an aggressor nation. A potential invader knows in advance that it will be difficult, if not impossible, to control the social and political life of that nation. On the one hand, civilian defense has the moral advantage that it can be used

3. Robert Woito, *To End War: A New Approach to International Conflict* (New York: Pilgrim Pr., 1982), p. 417.

only for defensive purposes. On the other hand, it does not forsake the moral obligation to defend the innocent. The claim of supporters of civilian defense is that it will deter aggression, like conventional and nuclear military deterrence, and if deterrence fails, it will avoid annihilation (possible with use of nuclear or even conventional weapons). Practically speaking, it seems as good as nuclear or conventional military deterrence, and morally it seems superior. The main advocates of civilian defense are Sharp (1970, 1973), Boserup and Mack, and Roberts. The bibliographies and references in these works will lead the interested reader to a body of literature hardly touched upon in this book.

Summation

As we have indicated, those who advocate living without war are termed pacifists. As Cady notes:

> Pacifism is regularly confused with passivism. "Pacifism" (from the Latin *pax, pacis,* peace, originally a compact + *facere,* to make) means, simply, peace making or agreement making. "Passivism" (from the Latin *passivus,* suffering) means being inert or inactive, suffering acceptance. Pacifist activists are committed to making peace, making compacts, making agreements, contributions to harmonious cooperative social conduct which is orderly by itself from within rather than ordered by the imposition of coercion from without.[4]

Of all the alternative futures considered in this chapter, only pacifism pursues peace without nuclear or conventional armaments; under civilian defense, it is a theory of nonmilitary deterrence among nations and a method for nonviolent resistance, if deterrence fails.

Conclusion

The dominant perspective today is that of living with nuclear weapons. As we have suggested, this perspective entails two quite different orientations: up until the early 1970s, living with nuclear weapons meant continued arms buildup to match the perceived intentions of the Soviet Union; a declaratory policy of assured destruction with a consistent force employment policy (SIOP); professed belief in the utility of arms control (but not disarmament); and a search for more accurate and efficient weapons. It meant a continuation of counterforce policy, while emphasizing (publically) countervalue targeting. In its place, a quite different orientation has arisen: one that emphasizes not parity, but superiority; prevailing in a conflict situation; distrust of arms control; and a desire to develop a defensive system that would reduce the vulnerability of exist-

---

4. Duane L. Cady, "Backing into Pacifism," *Philosophy and Social Criticism* 10 (1984): 173–74.

ing strategic forces. Targeting policy remains essentially the same—a combination of counterforce and countervalue options.

Both orientations envision a future wherein nuclear weapons continue to play a determining role in the search for a viable national security. While one orientation emphasizes deterrent capability, the other stresses victory, should deterrence fail. Both support measures reducing vulnerability to existing strategic forces, while differing strongly over the need for further weapon development. Politicians and strategists who were instrumental in the development of early policy and theories of deterrence and strategy now find themselves on the margins, but have become substantial critics of the more recently developed policies. While some have come to associate moral questions with the changes in policy, most remain supportive of a future with nuclear weapons, albeit with different policies concerning their use.

Citizens concerned with shifting policies and calls for further weapon development find themselves with difficult choices—either demanding a return to the earlier emphasis on a balance of terror or seeking alternatives not involving a dependence on nuclear weapons. Returning to a balance of terror emphasis would not eliminate the fears and anxieties over the outbreak of a nuclear war that have been recently manifested in protests and demonstrations, but would perhaps reduce the subjective probabilities associated with its occurrence. Seeking alternatives to nuclear dependence, while still emphasizing national security, is indeed risky, whether it be through a transition to conventional deterrence and military capability or toward political and diplomatic mechanisms that seek nonviolent conflict resolution. Persons advocating either of the latter perspectives are often ridiculed as being softheaded, impractical, or supporting destabilizing policies. Yet the voices of the latter perspectives have become a potent political force on the basis of their rejection of the legitimacy of a nuclear culture. These latter two groups criticize both the traditional and now-dominant wings of the living with nuclear weapons perspective, arguing that national security is being threatened, not furthered, through nuclear dependence. Religious groups, professional groups (such as Physicians for Social Responsibility), grass-roots organizations (such as Ground Zero), and others are demanding military and political change, either unilaterally or bilaterally. Their increasing sophistication is reflected in specific calls for such measures as no first use, a comprehensive test ban treaty, verifiable arms negotiations, arms reductions, and repudiation of space and counterforce weapons, civil defense, and the current emphasis on prevailing in a nuclear exchange.

There is little doubt that in the near future living with nuclear weapons will continue as a reality, but that seeking alternatives to the current reality will play a more important role in political debate and decision making in the long run. Thus, the debate over weapons and war has widened. What were, for decades, secret disagreements primarily among those of the strategic community and policymakers, have become matters of public debate—among not only politicians and military strategists but also philosophers, religious thinkers, academicians, and other interested citizens. That nuclear weapons and war itself have increasingly become matters of the public domain reflects both the increased concern with possible consequences and also the desire for democratic participation in the development and direction of policy and action.

Bibliography

The following bibliography includes references cited in the text and related suggested readings; references are organized according to the major divisions of the chapter. The full bibliographic reference for all sources cited in the text of this chapter may be found in either the annotated or unannotated bibliography corresponding to the division of the text in which the source is cited. Sources that are annotated are the most essential or significant.

Living with Nuclear Weapons

Brodie, Bernard. *War and Politics*. New York: Macmillan, 1973.

> The last book by Brodie. He continues his argument, begun in 1945 in *The Absolute Weapon,* for nuclear deterrence. He argues that nuclear weapons deter both nuclear and conventional war between the superpowers. It is the fear of escalation to large-scale nuclear war that makes nuclear deterrence prevent both types of superpower conflict.

Graham, Daniel O. *High Frontier: A New National Strategy*. Washington, D.C.: Heritage Foundation, 1982.

> Former deputy director of the CIA argues for various defensive systems to end MAD—technical means to stop missiles and protect the population. He argues for both ground and space-based antimissile systems and for extensive civil defense.

Gray, Colin. "War Fighting for Deterrence." *Journal of Strategic Studies* 7 (1984): 5–29.

> A recent statement of Gray's position that being able to defeat the Soviets in nuclear war is not sufficient. The United States must also aim to achieve victory at whatever level a conflict is fought.

Harvard Nuclear Study Group. *Living with Nuclear Weapons*. New York: Bantam, 1983.

> Widely cited statement of the position that nuclear weapons are here to stay. Though influenced by both schools of strategic thought, it sides more with the liberal school's support for arms control.

Jervis, Robert. *The Illogic of American Nuclear Strategy*. Ithaca, N.Y.: Cornell Univ. Pr., 1984.

> A careful analysis and sustained critique of countervailing strategy. In the tradition of Brodie, Jervis argues against the war-fighting school and for the war-deterring school, showing the dangers of regarding nuclear weap-

ons like conventional weapons and coupling U.S. strategic forces to NATO. He ends by favoring minimum deterrence.

McNamara, Robert. "The Military Role of Nuclear Weapons: Perceptions and Misperceptions." *Foreign Affairs* 62 (1983): 59–81.

One of McNamara's strongest statements in which he distances himself about as far from a war-fighting perspective as possible without totally rejecting nuclear weapons. He sees nuclear weapons to be useful only for deterrence.

National Conference of Catholic Bishops. "The Challenge of Peace: God's Promise and Our Response." *Origins* 13 (May 19, 1983): 1–32.

Much-publicized pastoral letter on war and peace in the nuclear age, prepared by the National Conference of Catholic Bishops. Considered by many to be the most important and influential contribution to the just war debate. They give only provisional sanction to nuclear weapons and only for deterrence (no first use is morally justified).

Union of Concerned Scientists. *The Fallacy of Star Wars.* New York: Vintage Books, 1984.

Detailed technical critique of the feasibility and sustained political critique of the stability of antisatellite systems and land, sea, and space-based ballistic missile defense systems.

Carter, Ashton B. and David N. Schwartz, eds. *Ballistic Missile Defense.* Washington, D.C.: Brookings, 1984.

Davis, Jacquelyn K., et al. *The Soviet Union and Ballistic Missile Defense.* Cambridge, Mass.: Institute for Foreign Policy Analysis, 1980.

Gompert, David C., et al. *Nuclear Weapons in World Politics: Alternatives for the Future.* New York: McGraw-Hill, 1977.

Stein, Jonathan B. *From H-Bomb to Star Wars: The Politics of Strategic Decision Making.* Lexington, Mass.: Heath, 1984.

Taylor, William J., Jr., Steven A. Maaranen, and Gerrit W. Gong, eds. *Strategic Responses to Conflict in the 1980s.* Washington, D.C.: Center for Strategic and International Studies, 1983.

Trout, Thomas B. and James F. Harf, eds. *National Security Affairs: Theoretical Perspectives and Contemporary Issues.* New Brunswick, N.J.: Transaction Books., 1982.

U.S. Congress. House Committee on Armed Services. Subcommittee on Research and Development. *Hearings on the Strategic Defense Initiative.* Washington, D.C.: Govt. Print. Off., 1984.

U.S. Department of Defense. *Annual Report, Fiscal Year 1982.* Washington, D.C.: Govt. Print. Off., 1981.

_____. *Annual Report, Fiscal Year 1984.* Washington, D.C.: Govt. Print. Off., 1983.

Living without Nuclear Weapons

Center for Defense Information. *Current Issues in U.S. Defense Policy.* New York: Praeger, 1976.

Through books, such as this one, and a newsletter, the Center for Defense Information argues for a strong military but against reliance on nuclear forces. An independent group composed largely of retired military personnel, the Center accepts money from neither government nor defense contractors. The Center analyzes U.S. military policies, weapon systems, and spending and develops alternatives when it thinks they are needed.

Dyson, Freeman. *Disturbing the Universe: A Life in Science.* New York: Harper, 1979.

Sees as a failure the pursuit of peace by the nuclear establishment and by those favoring world government and international organizations. Argues each nation needs to develop its own defense. He wants the elimination of all offensive weapons (which are "bad") and favors (as "good") such defensive weapons as anti-aircraft, anti-tank, and anti-ballistic missile systems.

Mearsheimer, John J. *Conventional Deterrence.* Ithaca, N.Y.: Cornell Univ. Pr., 1983.

Recent scholarly study of how conventional deterrence succeeds and fails. Considers the possibility of conventional deterrence at the tactical level. In addition to offering his own theory of conventional deterrence, Mearsheimer provides several carefully researched case studies.

Roberts, Adam. *Nations in Arms: The Theory and Practice of Territorial Defense.* New York: Praeger, 1976.

Argues that some European nations should move from nuclear deterrence to territorial defense. A national defense in depth approach would be defensive (only respond to foreign attack) and could use only conventional arms.

Alternative Defense Commission. *Defense without the Bomb.* London: Taylor & Francis, 1983.

Golden, James R. *Conventional Deterrence in NATO: Alternatives for European Defense.* Lexington, Mass.: Lexington Books, 1984.

Heilbrunn, Otto. *Conventional War in the Nuclear Age.* London: Allen & Unwin, 1965.

Heyer, Robert, ed. *Nuclear Disarmament: Key Statements of Popes, Bishops, Councils and Churches.* New York: Paulist Pr., 1982.

Holst, Johan J. and Uwe Nerlich, eds. *Beyond Nuclear Deterrence.* New York: Crane, Russak, 1977.

Kaufmann, William H. *Planning Conventional Forces, 1950–80.* Washington, D.C.: Brookings, 1982.

Komer, Robert. "Treating NATO's Self-Inflicted Wound." *Foreign Policy* 13 (Winter 1973–74): 34-48.

Quester, George H. *Deterrence before Hiroshima.* New York: Wiley, 1966.

Living without War

Bertz, Charles R. and Theodore Herman, eds. *Peace and War.* San Francisco: Freeman, 1973.

In addition to articles on "the war system," this anthology devotes even more attention to building a world peace system. Included are sections on world government, reforming the state system, and civilian defense.

Boserup, Anders and Andrew Mack. *War without Weapons.* London: Frances Pinter, 1974.

This text is focused on civilian defense and takes into account problems of repression and occupation. Even considers civilian defense as a form of external defense (deterrence and dissuasion).

Cousins, Norman. "Modern Man Is Obsolete." *The Saturday Review of Literature* (August 18, 1945): 5–9.

One of the first calls for world government in response to the atomic bombings of Hiroshima and Nagasaki. Foresees likelihood of proliferation and destabilization. Jacob Viner cites and rejects Cousins's argument when developing what was to become the establishment theory of nuclear deterrence. (Viner's position is discussed in Chapter 7.)

Ginsberg, Robert, ed. *The Critique of War: Contemporary Philosophical Explorations.* Chicago: Regnery, 1969.

In addition to several essays on alternatives to war, this anthology includes "A Bibliography of the Philosophy of War in the Atomic Age."

Mayer, Peter, ed. *The Pacifist Conscience.* New York: Holt, 1968.

An anthology on the history of pacifism from Lao-Tzu to Martin Luther King, Jr. Includes excerpts from such important figures as Kant, Thoreau, Tolstoy, James, Gandhi, Einstein, Freud, Niebuhr, Buber, Weil, Russell, and Camus. Has a helpful bibliography on war, pacifism, and nonviolence compiled and edited by William Miller. Its annotated "select list" of texts includes those books recommended by at least six members of a distinguished bibliography committee.

Roberts, Adam, ed. *Civilian Resistance as a National Defence.* Baltimore: Penguin, 1969.

An anthology that makes the case for civilian defense. Contains both historical and policy assessments by about a dozen leading advocates.

Sharp, Gene. *Exploring Nonviolent Alternatives.* Boston: Porter-Sargent, 1970.

Much briefer than his major work (see below), this book has succinct chapters on national defense without armaments and on civilian defense. The bibliography is especially helpful in that it is organized by topics.

_____. *The Politics of Non-Violent Action.* Boston: Sargent, 1973.

Major theoretical and practical text on nonviolent action. Covers many forms of protest, noncooperation, boycott, and intervention.

Weinberg, Arthur and Lila Weinberg, eds. *Instead of Violence: Writings by the Great Advocates of Peace and Nonviolence throughout History.* New York: Grossman, 1963.

Includes a section "After the Atomic Bomb (1946 to the present)." Goes back to "Before Guns (Sixth Century B.C.–1400 A.D.)."

Woito, Robert. *To End War: A New Approach to International Conflict.* New York: Pilgrim Pr., 1982.

Major text of World without War Council. Combines topical chapters and annotated bibliography. Especially relevant are chapters on "Social Change: Nonviolent Approaches" and "Peace Research."

Bedau, Hugo, ed. *Civil Disobedience: Theory and Practice.* New York: Pegasus, 1969.

Beres, Louis and Harry Targ, eds. *Planning Alternative World Futures: Values, Methods, and Models.* New York: Praeger, 1975.

Bondurant, Joan. *Conquest of Violence: The Gandhian Philosophy of Conflict.* Rev. Berkeley and Los Angeles: Univ. of California Pr., 1965.

Clark, Grenville and Louis Sohn. *Introduction to World Peace Through World Law.* Rev. Cambridge, Mass.: Harvard Univ. Pr., 1966.

Falk, Richard. *Nuclear Policy and World Order: Why Denuclearization.* New York: Institute for World Order, 1978.

———— and Saul Mendlovitz, eds. *The Strategy of World Order.* 4 vol. New York: Institute for World Order, 1966.

Fisher, Roger. *Points of Choice: International Crises and the Role of Law.* London: Oxford Univ. Pr., 1978.

Gandhi, M. K. *Non-Violent Resistance (Satyagraha).* New York: Schocken, 1951.

Gregg, Richard. *The Power of Nonviolence.* Rev. New York: Schocken, 1959.

Horsburgh, H. J. N. *Non-Violence and Aggression: A Study of Gandhi's Moral Equivalent of War.* London: Oxford Univ. Pr., 1968.

Johansen, Robert. *Toward a Dependable Peace: A Proposal for an Appropriate Security System.* New York: Institute for World Order, 1978.

King, Martin Luther, Jr. *Stride toward Freedom: The Montgomery Story.* New York: Harper, 1958.

Mendlovitz, Saul H., ed. *On the Creation of a Just World Order: Preferred Worlds for the 1990s.* New York: Free Pr., 1975.

Mullis, Walter and James Real. *The Abolition of War.* New York: Macmillan, 1963.

Osgood, Charles E. *An Alternative to War or Surrender.* Urbana, Ill.: Univ. of Illinois Pr., 1962.

Slyck, Philip Van. *Peace: The Control of National Power.* Boston: Beacon, 1963.

Somerville, John. *The Peace Revolution: Ethos and Social Process.* Westport, Conn.: Greenwood, 1975.

Stiehm, Judith. *Nonviolent Power.* Lexington, Mass.: Heath, 1972.

Wien, Barbara J., ed. *Peace and World Order Studies: A Curriculum Guide.* New York: World Policy Institute, 1984.

A Guide to Resources

This book outlines the major issues surrounding the nuclear arms race and the threat of nuclear war. Suggestions have been made for further reading on the topics. The literature cited, of course, represents only a small fraction of the available material. As concern over nuclear weapons and war has increased, high schools, colleges, and private religious and political groups have emphasized research in the area. Yet access to some of this material is often difficult, particularly to the material found in government publications and academic journals. This section provides a discussion of resource material, particularly reference works, that aid in that research process.

While college libraries vary widely in the scope and quality of their holdings, most have a large number of reference resources that help in gaining access to material and evaluating which works are necessary or important. Generally, reference works are either *fact tools*, or *finding tools*.[1] Fact tools contain specific information or data, including: (1) *yearbooks*—updates of material, such as an encyclopedia; (2) *handbooks and manuals*—current information on a topic (such as how to repair a car); (3) *almanacs*—collections relating to people, events, and so on (such as *Information Please Almanac*); and (4) *directories*—lists of people or organizations and descriptive information (such as address, officers) about them (for example, a phone book).

Finding tools, in contrast, refer to the location of sources. Three major finding tools are: (1) *card catalog*—a listing of works, usually books, by author, title, and subject; (2) *bibliographies*—listings of materials, usually by author and subject; (3) *indexes*—lists of sources of material such as magazines or journals by author and subject (for example, *Readers' Guide*

1. Anne K. Beaubien, Sharon A. Hogan, and Mary W. George, *Learning the Library. Concepts and Methods for Effective Bibliographic Instruction* (New York: Bowker, 1982), pp. 83–86.

to Periodical Literature). Some sources, such as encyclopedias, may be classified as either a fact or finding tool, depending on its use.

In this appendix, the selective emphasis is on those sources most likely to aid the person in gaining access to information on nuclear issues. Six resource areas are discussed: (1) the Library of Congress subject headings; (2) guides to bibliographic reference works; (3) indexes and abstracts; (4) academic journals; (5) government publications; and (6) other sources of periodic material such as handbooks and directory information. The intent here is to facilitate use of reference tools; they are indispensable in conducting original or secondary research (for example, term papers or theses), finding material for a lecture or background material for a study group or debate. To enhance the relevance of these materials, the accompanying descriptions are generally related to three subject areas that have formed the basis of this book: international relations, military affairs, and nuclear war and arms control.

The Library of Congress Subject Headings

In most cases, using the library is relatively straightforward. If you are interested in a book by an author (for example, Herman Kahn), you would turn to a card catalog or on-line computer system and look under the author heading "Kahn, Herman." If you are interested in one specific book Kahn wrote, say *On Thermonuclear War,* you would look under the title in the card catalog.[2] However, for books on thermonuclear war or nuclear war in general, use of the subject headings in the card catalog may result in few, if any, sources. This is because of the system used in cataloguing the library holdings.

Most college libraries use Library of Congress (LC) subject headings to index their collection. If you want to find books on a particular topic, you need to be able to identify the subject headings that the Library of Congress uses for the topic you've chosen. These subject headings, published periodically in book form and updated on microfiche, are entitled *Library of Congress Subject Headings;* each subject heading is listed in boldface type. Going back to the example above, let's say that you are interested in the topic of nuclear weapons and nuclear war. Prior to 1985, neither of these headings were used by the Library of Congress, as "atomic" was considered the generic term. Now, "nuclear" is considered the generic term. Table A1 reproduces these subject headings and related material in abbreviated form, and illustrates the major types of annotation:

---

2. If you know a specific author or book title, you can gain additional information using the tracings located on the bottom of the card. These tracings are subject headings assigned to the book and indicated by arabic numbers. You can use these subject headings to find other books on the same subject.

Table A1.
Library of Congress Subject Headings: Nuclear Weapons and Nuclear
Warfare

Subject Headings

Nuclear Weapons (Indirect) (U264)		Nuclear Warfare (U263)	
sa	Atomic bomb	sa	Art and nuclear warfare
	Hydrogen bomb		Communism and nuclear
	Neutron weapons		warfare
	No First Use (Nuclear Strategy)		Nuclear crisis control
	Nuclear Arms Control	X	Atomic warfare
___	Inventory control		CBR warfare
___	Testing	XX	Military art and science
___ ___	Detection		Naval art and science
	X Detection of nuclear		Nuclear weapons
	weapons tests		Nuclear war
		___	Religious aspects
		___	___ Baptists (Catholic Church,
			etc.)
		___	Social Aspects (Indirect)
			X Atomic warfare and society

Source: Library of Congress. Subject Cataloguing Division. *Library of Congress: Subject Headings Vol. II L–Z.* 10th ed. (Washington, D.C.: Library of Congress, 1986), pp. 2240–41.

1. To the right of a subject heading or one of its subdivisions (see point 5 below), you may find in parentheses a *class number* (or call number). For example, under the subject heading entitled "nuclear warfare," the class number U263 appears. Finding this call number in the book collection will lead to general material on this subject.

2. An *x* preceding a subdivision means that this and other terms following the *x* are not used by the LC subject heading system. For example, under the subject heading "nuclear weapons," an *x* appears beside the terms "atomic weapons" and "weapons, nuclear." This means that one will not find listings under those headings in the card catalog.

3. The *sa* abbreviation (*see also*) precedes additional subject headings listed under the major subject heading in boldface. For example, under the subject heading "Nuclear Weapons," *sa* refers to the terms "Atomic bomb" through "Nuclear Arms Control." The *sa* refers you to related subject headings that may be used in searching for materials.

4. An *xx* preceding a subject heading refers to a *broader* scope of subject headings than the one under examination. For example, under the subject heading "Nuclear Warfare" in Table A1, the *xx* annotation is associated with "Military art and science," "Naval art and science," "Nuclear war," and so on. The *xx* suggests that material on nuclear warfare may be found as a subdivision of the subject headings "Nuclear war" and "Military art and science." The relationship between the *sa* and *xx* annotations is reciprocal: by tracking the terms following the code

sa one should be led to either more specific or related terms. By tracking the terms following the code *xx* one should be led to broader terms.[3] When using the category "nuclear weapons," you may wish to become more specific (*sa*) by looking under the subject heading, "hydrogen bomb;" in contrast, if you decided to look under more general headings, or *xx*, you might look under "nuclear arms control."

5. The notation "-or-" refers to subdivisions creating more specific subject headings. For example, the terms "Social aspects" and "Religious aspects" under "Nuclear warfare" reflect a more specific subject heading.

Knowledge of these annotations aid in the effective use of the subject headings to find subject material in the card catalog. This is especially true in the area of nuclear issues, as the subject headings may differ substantially from the common vernacular. More generally, the *Library of Congress Subject Headings* provides a systematic method for the classification of books and other material. Indexes such as Wilson Company's *Readers' Guide* and *Social Science Index* often use the LC subject headings.

All too often, students give up searching for material if their initial use of the card catalog proves disappointing. *Ask for help!* Librarians can help in explaining the classification system, the reference collection, ways of approaching your topic, and so on. If you can't find an important source because it is not in the holdings, a librarian can often help you get a book or other material through an interlibrary loan system.

Guides to Bibliographic Reference Works

Most of the material listed or discussed in this section are secondary sources or finding tools—that is, they are compilations of original information (books, articles). By going from the general to the specific sources, such reference material can provide the basis for beginning research projects, understanding the various sides of debate, and, importantly, gaining access to a vast amount of information quickly, even when the actual material is not present in the library.

Listed below is a series of general reference aids. Each is bibliographic: it lists sources of other material. Major sources are annotated briefly; citations that follow are suggestive additional sources. The first group is oriented toward the popular press (newspapers, magazines). Access to these sources is helpful when one wants to create a chronology of events around a subject, or find current discussions and descriptions of a topic. Such sources normally would not constitute the basis for a

3. Library of Congress, *Library of Congress Subject Headings* 9th ed. (Washington, D.C.: Library of Congress, 1980), pp. ix–x.

research paper, but rather one used for anecdotal or illustrative purposes. Second are sources for bibliographies of bibliographies. This is perhaps the largest or most general level of reference work, helpful in the beginning stages of the research process. A third section describes a number of resource guides, distinguishing between guides to reference material, guides to literature in a field, and subject bibliographies.

Resources for Popular Literature

Alternative Press Index. An Index to Alternative and Radical Publications. Baltimore: Alternative Press Center, 1969–70.

There is a wide range of alternative periodicals that cover current events from a critical or leftist perspective. This index provides coverage of these newspapers and journals, and serves a useful function in listing sources and ideas that are often not covered in the traditional reference works.

Index to the Christian Science Monitor. Boston: Christian Science Monitor, 1960– .

The *Christian Science Monitor* has some of the best coverage of international relations and military affairs.

The National Newspaper Index. Los Altos, Calif.: Information Access, 1979– .

Indexes of the *New York Times, Wall Street Journal, Christian Science Monitor, Washington Post,* and *Los Angeles Times* are available in a combined microfilm format.

NYT Index. New York: The New York Times, 1851–1906/1912– .

This is perhaps the most widely used newspaper index.

The Readers' Guide to Periodical Literature. New York: Wilson, 1900–5– .

This is an index of popular magazines and other periodicals. Such a guide would be useful in researching a topic such as the recent push for a nuclear freeze.

The Vertical Index: A Subject and Title Index to Selected Pamphlet Materials. New York: Wilson, 1932– .

This is a useful guide to organizational literature, such as pamphlets, leaflets, and similar material.

Bibliographies of Bibliographies

Besterman, Theodore. *World Bibliography of Bibliographies.* 4th ed. 5 vols. Geneva, Switzerland: Societies Bibliographica, 1965.

While somewhat dated, this is a useful resource tool in gaining access to the vast international literature.

Bibliographic Index. New York: Wilson, 1937–42– .

This is an index of published bibliographies. The index lists sources in alphabetical subject headings; the sources may be books, articles, or separate publications. This can be a useful resource in developing reviews of literature.

Scull, Roberta A. *Bibliography of U.S. Government Bibliographies.* 2 vols. Ann Arbor, Mich.: Pierian, 1975–79.

This is an annotated listing of almost 2,000 bibliographies published by the federal government between 1968 and 1976. The arrangement is by discipline.

Guides to Reference Materials

Arkin, William M. *Research Guide to Current Military and Strategic Affairs.* Washington, D.C.: Institute for Policy Studies, 1981.

This is an informally written and easy to use research guide covering general information services, government documents, and serial publications. It is designed for "students, researchers, journalists, peace activists, government workers and others with a professional or scholarly interest in this field." One of the strongest aspects of this work is the delineation of journals and other serial publications. Journals on international relations, strategic and military topics, ground forces and weapons, naval forces, the Soviet Union, and arms control are listed, as well as newsletters on military posture and programs, military service, and information services on weapons. Serial publications of the U.S. government are also covered, along with information on the various congressional hearings and reports. This is the most useful guide in the area.

LaBarr, Dorothy F. and J. David Singer. *The Study of International Politics: A Guide to the Student, Teacher, and Researcher.* Santa Barbara, Calif.: ABC-Clio, 1976.

This guide provides primarily listings of contents of various works. The section on American and Comparative Foreign Policy covers a number of important works. The most relevant parts of this book cover reference material: journals and annuals (data-based, traditional, policy analysis), abstracts and book reviews, data sources and handbooks, and bibliographies.

McInnis, R. G. and J. W. Scott. *Social Science Research Handbook.* New York: Barnes & Noble, 1975.

This book lists and describes about 1,500 reference sources for the social sciences. The material is presented by discipline and geographical areas (such as Africa and Middle East; East Europe and the Soviet Union, etc.).

Sheehy, Eugene P. *Guide to Reference Books.* 10th ed. Chicago: American Library Assn., 1986.

This guide describes and analyzes research guides, bibliographies, abstracts, journals, indexes, sources of book reviews, dictionaries, handbooks, and directories within major subjects (for example, humanities, social sciences, history, and area studies). Subject, author, and title indexes are provided. Over 10,000 reference titles are covered.

Webb, William, ed. *Sources of Information in the Social Sciences.* 3rd ed. Chicago: American Library Assn., 1986.

Bibliographic and academic specialists describe the basic information sources in the social sciences (history, geography, economics and business administration, sociology, anthropology, psychology, and political science). Each discipline is covered substantively by a review of the basic works in the area, followed by a discussion of the basic reference and related sources (such as guides to literature, reviews of literature, abstracts and summaries, bibliographies, and journals).

Bertman, Martin A. *Research Guide in Philosophy.* Morristown, N.J.: General Learning Press, 1974.

Harmon, Robert B. *Political Science: A Bibliographic Guide to the Literature.* 3rd supplement. Metuchen, N.J.: Scarecrow, 1974.

McInnis, Raymond G. *Research Guide for Psychology.* Westport, Conn.: Greenwood, 1982.

McMillan, Patricia and James R. Kennedy, Jr. *Library Research Guide to Sociology. Illustrated Search Strategy and Sources.* Ann Arbor, Mich.: Pierian, 1981.

Stoffle, Carla and Simon Kartner. *Materials and Methods for History Research.* New York: Libraryworks, 1978.

Guides to Literature

Military and International Relations

Larson, Arthur D. *National Security Affairs: A Guide to Information Sources.* Detroit: Gale, 1973.

This is a bibliography covering almost 4,000 sources. The material is divided in seven parts; the following sections include relevant material: (1) International System; (2) Peace and Non-violence; (3) Arms and Arms Control; (4) Conflict and War; (5) Foreign Relations; (6) War and Strategy; (7) Defense Policy; (10) Conflict Resolution; (11) Games and Simulation; and (13) Bibliographies and Guides to Literature. Author and keyword indexes are provided. The vast majority of the literature covered is not annotated, but this remains a useful starting point to the literature.

Pfaltzgraff, Robert L. *The Study of International Relations: A Guide to Information Sources.* Detroit: Gale, 1977.

This is an annotated bibliography of classic and current works in the field of international relations. Delineation of topics such as deterrence, strategy, and use of power are covered in the chapters on foreign policy, military strategy, and theories of conflict. This is a particularly useful resource tool in understanding the different perspectives within the field of international relations.

Plischke, Elmer. *U.S. Foreign Relations: A Guide to Information Sources.* Detroit: Gale, 1980.

Four major topics are covered: (1) Diplomacy and diplomats; (2) Conduct of Foreign Relations; (3) Official Sources and Resources; and (4) Memoirs and Biographical Literature. Each chapter has a brief introduction to the sources. The individual chapters on the national security system and the military are of particular relevance, as is his analysis of documentation.

Bibliographies on the Military and International Relations

Amstutz, Mark R. *Economics and Foreign Policy: A Guide to Information Sources.* Detroit: Gale, 1977.

This is an extensive annotated bibliography covering such topics as theories of aid, trade, investments, and imperialism, primarily from an international political economy perspective. See particularly the chapter on the economics of war and defense.

Burns, Grant, ed. *The Atomic Papers: A Citizen's Guide to Selected Books and Articles on the Bomb, The Arms Race, Nuclear Power, the Peace Movement, and Related Issues.* Metuchen, N.J.: Scarecrow, 1984.

Over 1,100 articles and books focusing on the arms race (strategies, proliferation, and arms control), nuclear energy (pro and con), personalities (spies and atomic scientists), events (Cuban missile crisis, Three Mile Island), and art (fiction) on nuclear war. Popular literature is treated equally with scholarly literature. This is a useful compilation for research topics or ideas, rather than as a systematic treatment of the scholarly literature.

Burns, Richard Dean. *Arms Control and Disarmament: A Bibliography.* Santa Barbara, Calif.: ABC-Clio, 1977.

This is the sixth in the War/Peace Bibliography series under Burns's editorship. Over 8,000 sources are listed including both popular and scholarly works. In the Overviews and Theory section, reference resources are covered along with a series of substantive concerns (for example, UN, inspection and verification). The second section focuses on accords, proposals, and treaties concerning arms control (such as limiting weapons, controlling proliferation).

Christman, Calvin L. "Nuclear War and Arms Control." In Robin Higham and Donald J. Mrozek, eds., *A Guide to Sources of U.S. Military History, Supplement I.* Hamden, Conn.: Archon Books, 1981.

This is one of the only extensive bibliographic essays on nuclear issues. Christman provides a brief bibliographic overview from the early developments of the atomic bomb (and the personalities involved) changing military strategy, particular weapon systems, and issues of arms control. This is an excellent resource tool, both in the scope of issues and the quality of the sources.

Department of the Army. *National Security, Military Power and the Role of Force in International Relations. A Bibliographic Survey of Literature.* Washington, D.C.: Govt. Print. Off., 1976.

850 annotated sources (books, articles, proceedings, speeches) covering the 1946-76 period. Includes a substantial number of sources on Soviet strategies and policies, American foreign and defense policies, deterrence and arms control, and so on. This had good coverage of military reports and private military-oriented conferences.

Dextor, Byron, ed. Council on Foreign Relations. *The Foreign Affairs 50-Year Bibliography: New Evaluations of Significant Books on International Relations, 1920-1970.* New York: Bowker, 1972.

This is a compilation of books in the area of international relations that were deemed to be landmark or watershed works. The extensive annotations are written in an informal and readable manner. The works are divided into three subject areas: (1) general international relations; (2) the world since 1914; and (3) area studies. Chapter 10 of Part I and Chapters 3 and 4 of Part

II include references on the significance of nuclear weapons and their effect on military and international relations.

Groom, A. J. R. and C. R. Mitchell, eds. *International Relations Theory: A Critical Bibliography*. New York: Nichols, 1978.

This is a collection of bibliographic essays or review articles covering classic and current works in the area of international relations. It is particularly strong in its coverage of both American and European works. Four of the chapters (international stratification, conflict and war, foreign policy, order and change) examine material of relevance to an understanding of war in the nuclear age.

Higham, Robin, ed. *A Guide to the Sources of U.S. Military History*. Hamden, Conn.: Archon Books, 1975.

_____ and Donald J. Mrozek, eds. *A Guide to the Sources of U.S. Military History. Supplement I*. Hamden, Conn.: Archon Books, 1981.

The initial guide consists of nineteen review essays covering historical periods (such as American Revolution or Navy in the 19th Century), science and technology (in both the 19th and 20th centuries), recent writings on the army and navy (1945-73), and a chapter on the Department of Defense and Defense Policy. Each chapter has a section on general references, background surveys, policy, strategy, and tactics (where appropriate), and controversial topics. Supplement I updates the previous nineteen chapters for the 1973-78 period and includes a number of new chapters, two of which are relevant: one on nuclear war and arms control and a chapter on U.S. government documentation.

Roberts, Henry L., et al., eds. *The Foreign Affairs Bibliography*. New York: Bowker, 1973.

This is the fifth in the series of ten-year bibliographies, published under the auspices of the Council of Foreign Relations. It covers important works in the area during the 1962–72 period. While the annotations are generally brief, the scope of the material is a useful guide to the literature of the period. A fairly substantial number of works cited deal with the significance of the nuclear arms race and attempts at arms control.

Alexander, Susan, ed. *Bibliography of Nuclear Education Resources 1984*. Cambridge, Mass.: Educators for Social Responsibility, 1984.

American Security Council Education Foundation. *Quarterly Strategic Bibliography*. Washington, D.C.: ASCEF, 1977– .

Arms Control: A Bibliography with Abstracts. Springfield, Va.: National Technical Information Service, 1977.

Burns, Richard Dean. *Nuclear America: A Bibliography*. Santa Barbara, Calif.: ABC-Clio, 1984.

Carlson, Julia F. and Robert G. Bell. *Salt II: A Bibliography*. Library of Congress. Congressional Research Service Report 78-176F, 1978.

Disarmament: A Select Bibliography, 1973–77. New York: United Nations (Hammarskjold Library), 1978.

Greenwood, John. *American Defense Policy since 1945: A Preliminary Bibliography*. Lawrence, Kans.: Univ. Pr. of Kansas, 1973.

Lane, Jack C. *America's Military Past: A Guide to Information Sources.* Detroit: Gale, 1980.

Lang, Kurt. *Military Institutions and the Sociology of War: A Review of the Literature with Annotated Bibliography.* Beverly Hills, Calif.: Sage, 1972.

Petty, Geraldine, et al. *NEWS (Nuclear Energy, Weapons and Safeguards) Data Base: A Computerized Bibliography 1975–78.* Santa Monica, Calif.: Rand Corp., 1978.

Saltman, Juliet. "Economic Consequences of Disarmament." *Peace Research Reviews* IV 5(1972).

Sica, Geraldine P. *A Preliminary Bibliography of Studies of the Economic Effects of Defense Policies and Expenditures.* McLean, Va.: Research Analysis Corp., 1968.

United Nations. Institute for Disarmament Research. *Repertory of Disarmament Research.* New York: United Nations, 1982.

Wein, Barbara, ed. *Peace and World Order Studies: A Curriculum Guide.* 4th ed. New York: World Policy Institute, 1984.

Wills, Dennis. *Selected SALT Bibliography.* New York: Columbia Univ. Pr., 1977.

Indexes and Abstracts

While bibliographies may cover both books and articles, an index normally provides access to the periodical or serial literature (newspapers, magazines, journals). Moreover, bibliographies are often one-time publications; as such, they quickly go out of date. Indexes are continually being updated. The periodical literature is immense; for almost any topic, there may well be four or five times as much material available from periodicals as from published books. This is particularly true for scholarly works, and especially the analysis of nuclear-related issues. In any one year, there may be thirty or forty books and two hundred articles. And while college libraries expend large sums for periodical holdings, few are capable of carrying more than fraction of the available literature.

Indexes and abstracts help this situation by offering access, scope, efficiency, and currency. An index provides a kind of information retrieval system by making one aware of material classified by author, subject, title, and keywords (either a combination of words in the title or classifying the source by crucial ideas). The scope of material covered by most indexes is much greater than that of bibliographies, as they normally include related fields. The scope is also greater in that few libraries maintain periodical holdings as large as the coverage of an index. An index is efficient in the sense that one can gain quick access to a vast amount of material. This is true whether one is searching paper volumes manually or using on-line data bases. The development and spread of computer-based indexes have increased the speed and efficiency of searches may times over. Indexes also provide access to the most recently available material, as periodicals are more normally up to date than books.

Abstracting services are specialized types of indexes, providing summaries of the material. Such a service is crucial when one is faced with a hundred articles that seem to have the same title or focus; an abstract can help to select from sources without going to the original source. Further, the abstract can help one decide whether to initiate an interlibrary loan request. Taken together, abstracts and indexes facilitate both narrowing and broadening the sources for research. The indexes described below are those most likely to provide wide coverage of periodicals in the social services and humanities; more specifically, they have been found to be useful in seeking information on nuclear issues.

ABC Pol Sci: A Bibliography of Contents: Political Science and Government. Santa Barbara, Calif.: ABC-Clio, 1969– .
   A listing of tables of contents from selected journals in political science and government. The coverage is of American and various international periodicals. A keyword index facilitates compilation of bibliographies.
America: History and Life. Santa Barbara, Calif.: ABC-Clio, 1964– .
   This is a wide-ranging abstract and index service divided in four parts: (1) Article abstracts and citations (including article profiles based on descriptor terms); (2) Index to book reviews (including multiple reviews of the same book); (3) Bibliography of books, articles, and dissertations; and (4) Annual index of subjects, authors, book titles, and book reviewers. The rotated descriptors in the indexes aid in approaching a subject from a variety of perspectives. Nuclear issues are particularly covered in the categories of U.S. history 1945 to the present, and history, the humanities and social sciences.
Declassified Documents Reference System (DDRS). Woodbridge, Conn.: Research Publications, 1975– .
   Consists of a declassified quarterly catalog (with abstracts and subject indexes) and a retrospective collection of declassified documents that abstracts a large series of documents released in 1974. A microfiche copy of the documents is available with the system. While there are specific procedures to follow in seeking declassified documents, this resource stands out in its extensive abstracting of the material. Coverage includes such topics as national security, the Soviet Union, intelligence activities, and Vietnam. There are a substantial number of items relevant to nuclear strategy and policy.
Humanities Index. New York: Wilson, 1907/15–74– .
   This is an index to periodicals in the humanities, covering the periodical literature in the fields of archaeology, area studies, folklore, history, language, literary and political criticism, performing arts, philosophy, and religion. It provides cross-references and book reviews. This index is a particularly useful resource tool for persons wishing to focus on moral, religious, and philosophical perspectives as they relate to nuclear issues.

International Political Science Abstracts. Oxford, England: Basil Blackwell, 1951– .

This publication provides indexes and abstracts to the serial literature in political science and related areas. Abstracts are categorized into six subject areas, the most important of which (on nuclear issues) are the international relations and national and area studies. This is a particularly useful research guide to European perspectives on U.S.-U.S.S.R. relations, the arms race, nuclear proliferation and arms control. Subject and author indexes are provided.

Peace Research Abstracts Journal (PRAJ). Dundas, Ontario: Peace Research Institute, 1964– .

This is a monthly annotated abstract service. Works covered are categorized into subject areas such as arms limitation, international institutions, military situation, tension and conflict, nations and national policy, ideology, and international law.

Public Affairs Information Service Bulletin (PAIS). New York: Public Affairs Information Service, 1915– .

This publication provides coverage of articles, books, publications, U.S. government documents, and reports on a wide range of public policy. Coverage includes literature from social science, business, public administration, and education. Given its vast scope, it is a particularly useful resource for material not covered in other indexes (for example, reports of private organizations).

Social Sciences Citation Index (SSCI). Philadelphia: Institute for Scientific Information, 1969– .

A four-volume index covering journals, books, monographs, and proceedings in the social and behavioral sciences. This is a particularly useful tool to build bibliographies. Three related indexes are provided: (1) a citation index listing authors of articles or books and authors that have cited the works; (2) a source index providing bibliographic information for articles and other works listed in the citation index; (3) a "permuterm subject" index that creates subject categories by pairing significant words in the title of a referenced work. A fourth volume explains the scope and structure of the indexes and listing of source publications. The *SSCI* allows one to derive an indirect measure of the strength of one's work by the degree to which others refer to it. This is helpful in doing research on, for example, debates over nuclear weapons, strategy, or policy.

Sociological Abstracts. San Diego, Calif.: Sociological Abstracts, 1963– .

This abstracting service covers over 1,000 journals and serial publications in sociology and related social science fields. Coverage includes conference reports, panel discussions, book reviews, and scholarly articles. Particular coverage of nuclear issues is normally found in the sections on military sociology, social movements, and social change.

United States Political Science Documents (USPSD). Pittsburgh: University Center for International Studies, 1975– .

This publication provides indexes and extensive abstracts from over one hundred scholarly journals in the social sciences. The abstracts include keywords, tables, and citations. The subject index rotates topic areas, enhancing one's ability to find information from a variety of keyed phrases. Most of the important American periodical writings on nuclear war from a social science perspective are covered. Useful keywords to generate sources include arms race, nuclear weapons, military policy, nuclear proliferation, deterrence, and arms control. Given the quality of abstracts, this is a particularly useful resource tool for data from periodicals not held in many libraries.

Scholarly Journals

One of the basic types of coverage by most indexes is the scholarly journal. Scholarly journals contain professional analyses of various topics. They generally reflect more recent information than that found in books. The following list of journals represent a selective sample, based in part on their coverage of nuclear issues and international relations. Most are within the broad areas of social science and the humanities; other journals and periodicals could be included, such as those in physics or education. Most of the journals below are indexed by at least one of the sources described above.

American Academy of Political and Social Science. Annals. Philadelphia: Academy of Political and Social Science, 1899– . Indexed in ABCPolSci, Amer His, PAIS, SocSc.

This journal focuses on one subject per issue. Articles are scholarly and interdisciplinary. The book review section covers a wide range of works in the social sciences. This publication has focused periodically on military and nuclear issues; most recently this has included a 1983 (Vol. 469) examination of nuclear armament and disarmament and a 1981 volume on "National Security Policy" (Vol. 457).

Armed Forces and Society. Beverly Hills, Calif.: Sage, 1974– . Indexed in AmerHis.

An interdisciplinary journal focusing on military institutions, civil military relations, arms control and peacekeeping, and conflict management. This is a scholarly journal, characterized by theoretical discussions as well as case studies. It is a useful resource tool for those interested in the issue of the role of the military in modern society.

Bulletin of Peace Proposals. Irvington-on-Hudson, N.Y.: Universitetsforlaget, 1970– . Indexed in IntPolSc, Peace Research Abstracts.

A scholarly journal, interdisciplinary in orientation, that focuses on nonviolent mechanisms for conflict resolution. Original and reprinted reports and documents that focus on peace issues are included. This is a useful resource tool for analyses of current initiatives on peace, primary source for documents, and examinations of global militarization, militarism, and the problems and promises associated with arms control and disarmament.

Bulletin of the Atomic Scientists. Chicago: Educational Foundation for Nuclear Science, 1945– . Indexed in IntPolSc, SocAb.

This journal aims at both a professional audience and the general reader. While its overall aim is to discuss the role of the scientist in society, its predominant focus over the past few years has been on the threat of nuclear war and related issues. Debates are often presented (for example, on the viability of existing policy or proposed new weapon systems). The general message of the journal is that the arms race is bringing us closer to nuclear war. One of the strongest virtues of this periodical is that it brings together policymakers, academicians, and experts in weapon technology and arms control working toward reduced dependence on nuclear arms. This results in probably the widest range of topics of any easily accessible periodical, including those not normally covered in other journals (for example, nuclear fuel cycle, the policy of secrecy, technology transfers, civil defense). This magazine is an excellent resource tool for those doing papers in the social sciences, humanities and physical sciences.

Current Research on Peace and Violence. Tampere, Finland: Tampere Peace Research Institute, 1971– . Indexed in IntPolSc, Peace Research Abstracts, SocAb.

Formerly titled *Instant Research on Peace and Violence,* this quarterly emphasizes interdisciplinary scholarly work in the field of peace research. Analyses of issues are problem-oriented, rather than attempts at explanation (for example, model testing). This is a useful resource tool for the study of militarism, global militarization, and the social effects of armaments and disarmament.

Daedalus. Cambridge, Mass.: Norton Wood, 1955– . Indexed in PAIS, IntPolSc, Humanities.

This is the official publication of the American Academy of Arts and Sciences. Each volume examines a particular issue in the fields of education, science, politics, international relations, and policy analysis. Periodically, there are volumes examining military or strategic issues—see, for example, the fall 1980 and winter 1981 issues that examine U.S. defense policy.

Disarmament: A Periodic Review by the United Nations. New York: United Nations, 1978– . Indexed in PAIS.

This is a relatively new journal that focuses on a variety of peace issues. Articles include reports on United Nations activities, speeches, reports, and professional articles on disarmament. This is a useful resource tool for those interested in the different dimensions of disarmament, particularly initiatives taken by international organizations to reduce the threat of war and facilitate peaceful conflict resolution.

Foreign Affairs. New York: Council of Foreign Relations, Inc., 1922– . Indexed in ABCPolSci, AmerHis, IntPolSc, PAIS, USPSD.

This is probably the most prestigious journal in the field of international relations. Articles are written by experts in academia, policy analysis, and government service. Nuclear issues are generally examined from the perspective of political ramifications of policy and doctrine. Topics such as SALT, deterrence, arms control, and the nuclear freeze are covered as well as more general discussions of superpower relations or the role of nuclear weapons in NATO. This is a valuable resource tool not only in the scope of issues examined, but also the status of the writings. As the largest circulation journal, and one that is seen as essential by practitioners and theoreticians, its material has an important impact on nuclear policy and thought.

Foreign Policy. Washington, D.C.: Carnegie Endowment for International Peace, 1970– . Indexed in ABCPolSci, AmerHis, PAIS, SocSc.

This quarterly journal includes works by scholars and diplomats (and other practitioners). A wide range of foreign policy issues is examined, including nuclear issues. Recent volumes have, for example, included analyses of deterrence, the nuclear freeze debate, SALT II, and arms control. This is a useful research tool for persons doing research in social science and the humanities.

International Interactions: A Transnational Multidisciplinary Journal. New York: Gordon & Breach, 1974– . Indexed in PAIS, Social Science.

This quarterly scholarly journal focuses on a variety of international problems: models of world order, crisis analysis, negotiations and peaceful settlements, civil-military relations, and so on. The orientation is normally empirical and longitudinal. Recent volumes have examined patterns of war in the twentieth century (1982), arms transfers to the Third World (1983), pioneers in peace research (1983), and the risks of nuclear war (1984). This publication will be of use to persons researching the patterning and correlates of modern war, global militarization, and various empirical studies of the arms race.

International Journal. Toronto, Ont.: Canadian Institute of International Affairs, 1947– . Indexed in ABCPolSci, AmerHis, IntPolSc.

This is a Canadian quarterly focusing on international affairs. Each issue examines a single topic of political, military, diplomatic, or economic importance. Articles are scholarly in nature; recent issues have dealt with a number of nuclear-related concerns: Soviet foreign policy (1982), arms control (1981), and arms and doctrine (1978). While this is a useful resource tool in the general area of international relations, it is particularly useful in presenting the point of view of a nonnuclear nation.

International Organization. Madison, Wisc.: World Peace Foundation, Univ. of Wisconsin Pr., 1947– . Indexed in IntPolSc, AmerHis, PAIS, ABC-PolSci.

This is a multidisciplinary scholarly journal providing analyses of international or supranational organizations in global political and economic relations. Periodic coverage of issues such as arms control and nuclear proliferation makes this a useful resource tool, particularly in its emphasis on international political economy.

International Security. Cambridge, Mass.: MIT Pr., 1976– . Indexed in ABC-PolSci, AmerHis, IntPolSc, PAIS.

This is one of the two or three most important journals in the understanding of nuclear issues. Scholarly presentations on policy, strategy, and historical developments are found in almost every issue. Recent articles have focused on changing nuclear strategy, nuclear proliferations, deterrence, weapon vulnerability, and the NATO alliance. This is an excellent resource tool for scholarly work in foreign affairs, international relations, history, strategic studies, and policy analysis.

International Social Science Journal. Paris: UNESCO, 1949– . Indexed in PAIS, SocSc.

A scholarly journal, with each issue focusing on a single topic. Nuclear-related concerns are covered with some frequency—see, for example, the 1983 issue on militarization in developed countries. This is a useful resource for research, particularly for persons interested in a multidisciplinary approach to global problems.

International Studies Quarterly. Beverly Hills, Calif.: Sage, 1957– . Formerly *Background on World Politics* and *Background: Journal of the International Studies Association.* Indexed in ABCPolSc, IntPolSc, SocSc.

This is the official journal of the International Studies Association. It has an interdisciplinary and international perspective. Articles are scholarly and empirical. Recent issues have included articles on deterrence, the arms race, and nuclear proliferation. This is a useful resource tool for advanced undergraduate courses in political science, and crosscultural analyses of economic and political global problems.

Journal of Conflict Resolution. Beverly Hills, Calif.: Journal of Conflict Resolution, 1957– . Indexed in PAIS, Peace Research Ab, SocSc.

This interdisciplinary quarterly focuses on international conflict, with a heavy emphasis on theoretical and empirical models. Much of the important work examining the arms race has appeared in this journal. Extensive review articles provide important source material on a variety of subjects relevant to the arms race and global militarization.

Journal of Peace Research. Irvington-on-Hudson, N.Y.: International Peace Research Institute, 1964– . Indexed in IntPolSc, PAIS, Peace Research Abstracts.

This is a scholarly journal emphasizing ''articles directed toward ways and means of promoting peace.'' Recent issues included articles on the etiology of war, militarization, disarmament, and effects of the military on the Third World. This is a useful periodical for its international perspective and consistent critique of the arms race.

Journal of Peace Science. Binghamton, N.Y.: State University of New York, 1973– . Indexed in ABCPolSci, Peace Research Abstracts.

Scholarly and professional journal focusing on international relations. Theoretical and empirical in orientation, it is a useful publication for those interested in material on arms control and disarmament, nonviolent conflict resolution, patterns of war, and so on.

Orbis: A Journal of World Affairs. Philadelphia: Foreign Policy Research Institute, 1957– . Indexed in ABCPolSci, AmerHis, IntPolSc, PAIS.

Emerging as one of the more important journals in foreign policy, this publication has frequent articles related to nuclear arms and military policy. Scholarly in orientation, recent volumes have included articles on NATO and nuclear policy, arms trade, arms control negotiations, detente, U.S.-U.S.S.R. relations, and general examinations of nuclear strategy and war. The book review section is of particular value to the researcher. This is a useful periodical for those interested in strategic studies, policy analysis, the arms race, and analyses of policy and doctrine by important ''insiders'' (former or future nuclear policymakers).

Parameters: Journal of the U.S. Army War College. Carlisle Barracks, Penn.: U.S. Army War College, 1971– . Indexed in ABCPolSci.

This journal focuses on a wide variety of military and foreign policy issues. It includes both academic and military writings. Recent articles included examinations of the morality of theater nuclear weapons, Presidential Directive 59, and the implications of new weapon technology for defense policy.

Peace Research Reviews. Dundas, Ontario: Peace Research Institute, 1967– . Indexed in ABCPolSci, Peace Research Abstracts.

Each issue of this periodical focuses on a specific topic from an interdisciplinary perspective (natural and social sciences, and humanities). This is a useful publication for those interested in alternative models of national security and international relations. The bibliographies are also of use in the generation of sources.

Policy Studies Journal. Urbana, Ill.: University of Illinois, 1972– . Indexed in ABCPolSci, IntPolSc, PAIS.

The focus of this scholarly publication is on the efficacy of government programs to deal with social problems. Alternate issues focus on a single topic. The book reviews and review essays are of particular use for research purposes. Recent volumes have included an examination of American security policies, military strategic doctrine, and an extensive bibliography on national security affairs.

Political Science Quarterly. New York: Academy of Political Science, 1886– . Indexed in ABCPolSci, AmerHis, PAIS, SocSc.

This is a scholarly quarterly dealing with a wide range of political issues at the national and international levels. There are frequent articles of interest to the researcher of nuclear issues, particularly material on the Soviet Union, nuclear strategy and doctrine, and the political significance of specific weapon systems. Recent articles have examined theater nuclear weapons in Europe, forward-based weapons, the Soviet view of MX, origins of massive retaliation, space weapons, and deterrence of the Soviet Union.

Survival. London: International Institute for Strategic Studies, 1959– .

A publication of the IISS, one of the major independent sources of data and analysis on world military issues. This journal includes scholarly articles, important documents and reprints. The focus is on international peaceful conflict resolution. This is a particularly useful resource tool for persons interested in strategic studies, arms control and disarmament, foreign policy, and the role of the military in international relations.

World Politics: A Quarterly Journal of International Relations. Princeton, N.J.: Center of International Studies, Princeton Univ. Pr., 1948– . Indexed in ABCPolSci, AmerHis, InterPolSci, SocSc.

This journal provides excellent historical and current analyses of global issues. Scholarly in orientation and supported by detailed bibliographies and review essays, this journal is a useful resource tool for those interested in strategic studies, the arms race, regional and international political regimes. Recent articles have examined proposals for reducing strategic arms, current countervailing strategy, theater nuclear weapons, Soviet images of the U.S. nuclear policy, and just war theory.

Government Publications

The United States government publishes a large number of periodic reports, studies, and hearings. Access to this body of literature is difficult for a number of reasons. First, government publications are normally not part of library card catalogs or similar on-line systems. Second, these publications are not generally referenced in the various types of works listing books in print. Third, government publications are often kept in a separate section of the library, distinct from periodicals and books. Fourth, some libraries, designated as depository libraries, receive large numbers of works in different formats (book, microfilm, maps, and so on), while others have little more than general reference works on the publications (but not the publications themselves). And fifth, most libraries with large government holdings depend on a classification system (termed *SuDoc,* for Superintendent of Documents) that differs from the Library of Congress system.[4]

Given these difficulties, it is perhaps not surprising that many students do not use government publications in their research. Yet, this represents a major source of information and analysis. In the first section, some of the major guides and indexes to government publications are described. In the second section, attention is given to sources of publications of the executive branch. General guides and a selective bibliography of reports and documents are included. The third section focuses on congressional publications. Descriptions of basic publications, indexes, and selective agencies serving Congress are provided.

General Guides to Government Publications

The sources listed below provide useful descriptions and analyses of government publications: accessibility, availability, scope, types of materials, and so on. The Andriot work is different in that it delineates the various publications and provides indexes by author, agency, and SuDoc number. The other books provide detailed information on the various secondary reference tools (indexes, bibliographies) essential in accessing much of the literature.

Andriot, John L. *Guide to U.S. Government Publications.* Vol. 1. Arlington, Va.: Documents Index, 1982.
Herman, Edward. *Locating U.S. Government Information.* Buffalo, N.Y.: William Hein, 1983.
Morehead, Joe. *Introduction to U.S. Public Documents.* 3rd ed. Littleton, Colo.: Libraries Unlimited, 1983.

---

4. For the past few years, the subject index has used the Library of Congress Subject Headings. See Superintendent of Documents, *Monthly Catalog of United States Government Publications. Cumulative Index 1984* (Washington, D.C.: Govt. Print. Off., 1984), prelim. p. 10.

Schmeckebeir, Laurence F. and Roy B. Eastin. *Government Publications and Their Use.* 2nd rev. ed. Washington, D.C.: Brookings, 1969.

General Government Indexes

There are a number of basic reference works with which one should be familiar when approaching government publications. These works generally facilitate access to basic documents, follow legislation, or locate legislative reports and studies. Each of these indexes is accessible through bound volumes and on-line data base systems such as DIALOG and ORBIT.

American Statistics Index Annual (ASI). Washington, D.C.: Congressional Information Services, 1973– .

This is a monthly and cumulative annual publication focusing on publications that contain government statistics. Arrangement in each issue is by the government agency issuing the document. Abstracts are provided, as well as name, subject, title, and category indexes. This index is particularly useful to persons wishing to bring together material on the same subject from a number of different sources (for example, Defense Department vs. congressional analyses of military expenditures).

Congressional Information Service Index/Abstracts. Washington, D.C.: CIS, 1970– .

Comprehensive index and abstracting service of congressional documents, reports, committee prints, hearings and special publications. Published monthly with annual cumulations. The information is arranged first by congressional committee; within each committee, publications are listed by type. The main index covers subjects of each part of a hearing, and identification of persons giving testimony. Two cumulative annual indexes have been released, one for 1970–79 and the other for the 1975–78 period. Material relevant to nuclear issues is regularly included in both congressional hearings and reports (for example, studies conducted for Congress by the Congressional Research Service).

The Federal Index. Cleveland, Ohio: Predicast, 1976– .

This is a monthly index (with cumulative annual) of documents dealing with the federal government including *Congressional Record, Federal Register, Weekly Compilation of Presidential Documents, Commerce Business Daily, Law Week,* and *Washington Post.*

Government Reports Announcements and Index. Washington, D.C.: Commerce Department, 1975– .

Often termed the NTIS index (National Technical Information Service), this publication covers reports written as part of research contracts. Broad subject categories are provided and the indexes are issued bimonthly and cumulated annually. There are separate indexes for personal authors and corporate sources, and a subject index generated by use of title keywords. This series of volumes contains information on many studies not covered in other government indexes, and although somewhat technical, represents a rich source of material in the areas of nuclear weapons and nuclear strategy.

Monthly Catalog of U.S. Government Publications. Washington, D.C.: Govt.
  Print. Off., 1895– .
    Issued by the superintendent of documents, the *Monthly Catalog* lists all
  publications of the government on a monthly, semiannual, and annual basis.
  Information includes author, title imprint (who, when, and where published), Library of Congress subject heading, and series title. This is perhaps the primary source of information when first approaching the government literature. There have been decennial cumulative indexes for 1941–50
  and 1951–60, and more recent cumulated indexes for the 1961–65,
  1966–70, and 1971–76 periods. Additionally, there has been a cumulative
  subject index for the 1900–71 period and a title index for the 1789–1976
  period.

Sources of Publications of the Executive Branch

Three general types of publications are listed below. First are the
various executive branch papers—this includes the works of the various
presidential advisory groups, the Office of Management and Budget, the
Central Intelligence Agency, the *Federal Register*, and the two most often used reference sources for executive activity, the *Weekly Compilation of Presidential Documents* and the *Public Papers of the President.*
The second group of sources includes the annual reports or statements of
cabinet secretaries or agency heads. These sources are particularly helpful in researching changes in policy and postures of administration officials with reference to national security affairs. The third category of
works consists of publications or guides to treaties and other types of
international agreements.

Guides to and Publications of the Executive Branch

The Budget of the United States Government. Washington, D.C.: Govt. Print.
  Off., 1972– .
    The budget provides a series of proposals and supporting rationales on
  an annual basis. Supporting documents include an *Appendix, Special Analyses,* and *Budget Revisions.* The documentation on military spending is particularly useful as a resource tool.
Department of State Bulletin. Washington, D.C.: Govt. Print. Off., 1939– .
    Issued monthly, the *Bulletin* represents the official record of U.S. for-
  eign policy. It includes presidential statements and documents relating to
  foreign policy, statements and testimony given by officials of the State Department, and information on treaties or other agreements involving the
  United States.
The Federal Register. National Archives and Record Service. Office of the Federal Register. Washington, D.C.: Govt. Print. Off., 1936– .
    This publication includes regulations and notices of federal agencies as
  well as presidential executive orders and proclamations. A special edition of
  the *Federal Register* is the *Weekly Compilation of Presidential Documents,*
  which includes texts of proclamations, speeches, letters to Congress, an-

nouncements, and a list of press releases. The *Weekly Compilation* is cumulated in the annual series entitled *Public Papers of the President.*

Korman, Richard I., comp. *Guide to Presidential Advisory Commissions, 1973–1981.* Westport, Conn.: Meckler, 1982.

This guide includes a chronology of commissions, lists of members, and an abstracted delineation of reports issued by the commissions.

Sullivan, Linda E. *The Encyclopedia of Governmental Advisory Organizations.* 3rd ed. Detroit: Gale, 1980.

This guide provides general information on a wide variety of advisory committees including presidential advisory, public advisory, and interagency committees. Of particular use are the categories of popular names of the committee or its reports. For example, an advisory commission was established a few years ago to study basing the MX missile; the committee was popularly known as the Scowcroft Commission and its report the Scowcroft Report after its chairman. Often committees are better known by popular names of chairpersons; reference guides such as this help gain access to the reports.

Annual Reports or Statements

Department of Defense. *Department of Defense Annual Report.* Washington, D.C.: Govt. Print. Off., 1980– .

————. *Organization of the Joint Chiefs of Staff. United States Military Posture for Fiscal Year 19—.* Washington, D.C.: Govt. Print. Off., 1983– .

Department of Energy. *Department of Energy Annual Report.* Washington, D.C.: Govt. Print. Off., 19—– .

Department of State. U.S. Foreign Policy: A Report of the Secretary of State. Washington, D.C.: Govt. Print. Off., 19—– .

U.S. Arms Control and Disarmament Agency. *ACDA Annual Report.* Washington, D.C.: Govt. Print. Off., 19—– .

Treaties and International Agreements

Bevans, Charles I., comp. *Treaties and Other International Agreements of the U.S.A., 1776–1949.* 13 vols. Washington, D.C.: Govt. Print. Off., 1969– .

Treaties and Other International Acts Series (TIAS). Washington, D.C.: Dept. of State, 1946– .

U.S. Department of State. *Treaties in Force.* Washington, D.C.: Govt. Print. Off., 1956– .

————. *United States Treaties and Other International Agreements (UST).* Washington, D.C.: Govt. Print. Off., 1952– .

Related Documents

American Foreign Policy, 1950–1955: Basic Documents. 2 vols. Washington, D.C.: Govt. Print. Off., 1957.

American Foreign Policy: Current Documents, 1956–1967. 12 vols. Washington, D.C.: Govt. Print. Off., 1959–1969.

A Decade of American Foreign Policy: Basic Documents, 1941–49. Prepared by the Department of State. Senate Document 123, 81st Congress, 1st sess. Washington, D.C.: Govt. Print. Off., 1950.

Grenville, J. A. S. *The Major International Treaties, 1914–1973: A History and Guide with Texts.* New York: Stein & Day, 1974.

Schlesinger, Arthur M., Jr., ed. *The Dynamics of World Military Power: A Documentary History of U.S. Foreign Policy 1945–1973.* 5 vols. New York: Chelsea House, McGraw-Hill, 1973.

U.S. Arms Control and Disarmament Agency. *Arms Control and Disarmament Agreement: Texts and History of Negotiations.* Washington, D.C.: Govt. Print. Off., 1975.

_____. *Documents on Disarmament, 1961–* . Washington, D.C.: Govt. Print. Off., 1961– .

Congressional Publications

In general, congressional materials fall into two categories. The first consists of publications of the Congress as a whole, or by the House or Senate. Included in this category are the *Congressional Record*, the *Journals* of the House and Senate, the *Calendars* of the House and Senate, bills, joint resolutions, concurrent resolutions, and *Resolutions* of the House or the Senate. The second category consists of publications of the committees or subcommittees. Calendars, directories, hearings, committee prints, and reports and documents are examples of such publications.

Table A2 lists some of the major types of publications with brief explanations of their scope, and a partial indication of index sources. The types chosen are those most likely to have relevance for military and foreign affairs. Tables A3 and A4 list the major committees and subcommittes in both the House and Senate that have responsibility for military matters or foreign affairs. Also included in these charts are annual or periodic hearings of these committees—a rich source of information on current and proposed plans affecting the military. The works by Schmeckebier and Eastin, Morehead, and Herman cited at the beginning of this section should be consulted for more detailed discussions of the various publications and sources of reference material pertaining to Congress. For our purposes, three related types of resource tools briefly can be considered: (1) general publications; (2) the various indexes of congressional publications; and (3) the works of agencies that supply reports and research to Congress (CBO, GAO, CRS).

General Congressional Publications

Congressional Record. Washington, D.C.: Govt. Print. Off., 1873– .
    Published both in a daily edition and subsequent bound edition, the *Record* is the major resource tool for information on congressional legislation.

Table A2.
Selected Types of Congressional Publications

Type	Scope	Indexed
Bills	Proposals for enactment, amending existing laws.	CIS Index, CQ Weekly Report, Congressional Report
Committee prints	Reports by Congressional Research Service or committee staff.	Monthly Catalog, CIS Index
Hearings	Verbatim transcript of meetings and hearings about pending legislation, topics committees have authorization to investigate or oversee. Additional material often included.	Monthly Catalog, CIS Index
Reports	Statements describing findings and recommendations submitted by committee (includes analyses and minority opinions).	Monthly Catalog, CIS Index
Congressional record	Proceedings and debates. Includes speeches, articles, voting data, inserted material.	Monthly Catalog, Congressional Record Index
Documents	Presidential messages to Congress; executive agency statements to Congress.	Monthly Catalog, CIS
Senate executive documents	Presidential messages to Senate; may contain texts of proposed treaties and supporting materials.	Monthly Catalog, CIS

Source: Adapted from Edward Herman, *Locating U.S. Government Information* (Buffalo, N.Y.: William Hein, 1983), pp. 44–48.

In the daily edition, the proceedings of each chamber are provided, along with "Extensions of Remarks" and a "Daily Digest." The bound volumes include a cumulated index and "Daily Digest." In using this document, one should be fully aware of the material categorized as extension of remarks. One estimate is that fully 70 percent of the material in the *Record* is never actually part of the verbal transactions; a sizable proportion of this extra material falls in the extension category.

United States Congressional Serial Set. Washington, D.C.: Govt. Print. Off., 1817– .

At the end of a congressional session, reports and documents (see Table A2) are compiled and issued as volumes in the *Serial Set.* The *Set* covers the period from 1817 to the present. A retrospective series covering the first fourteen Congresses (1789–1814) entitled *The American State Papers* (Washington, D.C.: Gales and Seaton, 1832–61) was also issued. Reports describe the purpose and intent of legislation. Supporting material may include budget projections or analyses of the impact of proposed legislation. Special reports are also included (such as investigative activities of a com-

Table A3.
U.S. House Committees and Subcommittees with Responsibility for Strategic and Military Affairs

	Committee on Appropriations	Committee on Armed Service	Committee on Foreign Affairs	Committee on Budget	Permanent Select Committee on Intelligence
Subcommittees	Defense Foreign Affairs Military Construction State, Justice, Commerce, Judiciary	Procurement and Military Nuclear Systems Seapower and Strategic and Critical Materials Research and Development Military Personnel Military Installations and Facilities Military Compensation Special Subcommittee on NATO Standardization Readiness	International Security and Scientific Affairs Europe and Middle East Asian Pacific Affairs International Economic Policy and Trade Inter-American Affairs Africa International Organizations	Defense and International Affairs	
Hearings	Defense Appropriations Military Construction Authorization for Appropriations Foreign Assistance	Military Posture and Department of Defense Authorization Military Construction Authorization Department of Energy National Security and Military Application of Nuclear Energy Authorization	Foreign Assistance Legislation	Overview of Budget for Defense and International Affairs	

Source: Adapted from William M. Arkin, *Research Guide to Current Military and Strategic Affairs* (Washington, D.C.: Institute for Policy Studies, 1981), pp. 35–36, 38.

Table A4.
U.S. Senate Committees and Subcommittees with Responsibility for Strategic and Military Affairs

	Committee on Appropriations	Committee on Armed Services	Committee on Foreign Relations	Select Committee on Intelligence	Committee on the Budget
Subcommittees	Defense Military Construction Foreign Operations State, Justice, Commerce, Judiciary	Arms Control General Procurement Manpower and Personnel Research Development Military Construction and Stockpiles Procurement Policies and Reprogramming	International Economic Policy Arms Control, Oceans International Operations and Environment African Affairs European Affairs East Asian Pacific Near East and South Asian Affairs Western Hemisphere Affairs	Intelligence and Rights of Americans Budget Authorization Collection and Development Charters and Guidelines	
Hearings	Department of Defense Appropriations FY- Military Construction Appropriations FY- Foreign Assistance and Related Programs Appropriations FY-	Department of Defense Authorizations for Appropriations for FY- (Military Procurement, Research and Development, Active Duty, Selected Reserve Civilian Personnel Strengths)	Foreign Relations Authorization FY- International Development Assistance Authorization FY- International Security Assistance Programs FY-		First Concurrent Resolution on the Budget, FY- (Defense)

Source: Adapted from William M. Arkin, *Research Guide to Current Military and Strategic Affairs* (Washington, D.C.: Institute for Policy Studies, 1981), pp. 35–36, 38.

mittee or legislative histories) in the *Set*. The documents included in the *Set* are primarily presidential communications to the Congress. It is the reports section of the *Set* that is most useful in researching military and national security issues as it includes material on legislative intent.

United States Code. House of Representatives. Office of the Law Revision Counsel. Washington, D.C.: Govt. Print. Off., 1926– . Published every six years. Annual supplements.

The *Code* classifies federal law in forty-nine categories, three of which have particular relevance here: no. 10 (armed forces), no. 22 (foreign relations and intercourse), and no. 50 (war and national defense). There is a general subject index, a popular name index, and tables delineating presidential documents.

Indexes to Congressional Activity

CIS Index to Publications of the United States Congress. Washington, D.C.: Congressional Information Services, 1970– . (See General Government Indexes above.)

CIS United States Congressional Committee Hearings Index. Washington, D.C.: Congressional Information Service, 1981– . 8 parts.

This index supersedes the previously published *Index of Congressional Committee Hearings (Not Confidential in Character)*. It is a retrospective index covering the period from the early 1800s to 1969. There is a reference bibliography section and subject, title, report, bill number, and witnesses indexes.

CIS United States Congressional Prints Index from the Earliest Publications through 1969. Washington, D.C.: Congressional Information Services, 1980. 5 vols.

Similar to the *Hearings Index*, the *Prints Index* provides a research bibliography and a series of indexes: subject, title, committee, bill numbers, and so on. This research tool is valuable for its coverage of reports conducted by the Congressional Research Service and various committee staffs.

CIS United States Serial Set Index. Washington, D.C.: Congressional Information Service, 1975–79. 12 parts.

This index covers the period 1789 to 1969. Indexes are provided by subject and keywords, names of individuals and organizations, and a list of reports and documents.

Congressional Index Service. Chicago: Commerce Clearinghouse, 1937– .

This is a private, commercial index that is useful in deriving legislative histories. Subject and author indexes are provided. Background material on current legislation is provided, as well as summaries and status reports on legislation.

Agencies Serving Congress

Congressional Budget Office (CBO)

The Congressional Budget Office "provides Congress with basic budget data and with analyses of alternative fiscal, budgetary, and programmatic policy issues.'' Of the five publication series available to depository libraries, two, the *Background Papers* and the *Budget Issues Papers*, have particular relevance in researching nuclear and more general military issues. For example, recent analyses submitted to Congress covered such topics as MX missile basing, deterrence policies and procurement, European theater nuclear forces, and issues involving strategic nuclear forces.

Congressional Research Service (CRS)

Part of the Library of Congress, the CRS functions to provide Congress with reference material and general information. *CRS Studies in the Public Domain* lists current works. Other works are indexed in the *Monthly Catalog* and the *CIS Index*. Most CRS reports are issued by congressional committees and subcommittees as prints. On a regular basis, it presents analyses of military spending *(Defense Budget: FY—)*, foreign policy *(Congress and Foreign Policy, 19—)*, foreign relations *(Legislation on Foreign Relations through 19—, Vol. I-II: Current Legislation and Related Executive Orders, Vol. III: Treaties and Related Material)*, and arms control *(Evaluation of Fiscal Year 19—. Arms Control Impact Statements)*. It also produces a wide series of nuclear related reports. Recent topics have included arms control and weapon modernization (1983), U.S.-U.S.S.R. strategic nuclear forces (1984), and conceptual frameworks and strategic issues (1982). Gaining information on CRS reports should include use of the *Monthly Catalog, CIS Index*, or the *Library of Congress Information Bulletin*.

Government Accounting Office (GAO)

The GAO is an independent agency advising Congress on the use of public funds. The comptroller general issues an *Annual Report* and a *Publications List;* the GAO reports to Congress, particularly those of the science and technology subdivision, offer relevant material on nuclear and strategic issues. This subdivision reviews military (including nuclear) research and development and monitors the status of weapon systems. Use of the *Monthly Catalog* and *CIS Index* facilitates access to the GAO reports.

Office of Technology Assessment (OTA)

The OTA was established in 1972 to analyze impacts of existing or proposed technology and to identify alternative technological methods and programs. A list of reports conducted by OTA is included in its *Annual Report* to Congress. In recent years, it has conducted a number of nuclear related studies, including *The Effects of Nuclear War* (1979) and recent analyses of arms control in space (1984) and MX missile basing (1981).

Other Periodic Sources

In addition to a wide number of government publications dealing with the military and specifically with nuclear issues, there are a number of other sources that are useful in doing research. This section lists a number of publications sources, focusing on those of a recurring nature. Four types of publications are described: (1) handbooks, yearbooks, and annual surveys relevant to the military; (2) foundations and research centers that produce ongoing analyses of military and national security issues, either through journals or reports; (3) peace studies; (4) publications of the United Nations.

Handbooks, Yearbooks, Recurring Analyses

Banks, Arthur S. *Political Handbook of the World 1976–* . New York: McGraw-Hill, 1927– .

Copley, George R., ed. *The Defense and Foreign Affairs Handbook 1977–* . New York: Franklin Watts, 1978.

Defense Marketing Service, Inc. *DMS Market Intelligence Report: Foreign Military Market*. Greenwich, Conn.: DMS, Inc.

Dupuy, Trevor N., et al., eds. *The Almanac of World Military Power*. 4th ed. New York: Bowker, 1979.

Facts on File: World News Digest with Cumulative Index. New York: Facts on File, 1940– .

Foss, Christopher F., ed. *Jane's Armour and Artillery*. London: Jane's Publishing Co., 1979.

Hayes, Grace, ed. *Almanac of World Military Power*. Novato, Calif.: Presidio, 1970.

Hogg, Ian V., ed. *Jane's Infantry Weapons*. London: Jane's Publishing Co., 1975.

International Institute for Strategic Studies. *Military Balance, 1964–* . London: IISS, 1964.

_____. *Strategic Survey, 1964–* . London: IISS, 19—.

Nuclear Weapons Data File. London: Aviation Studies Atlantic, 1979– .

Pretty, Ronald T., ed. *Jane's Weapon Systems*. London: Jane's Publishing Co., 1969.

Royal United Services Institute for Defence Studies. *R.U.S.I. and Brassey's Defence Yearbook 1886–* . New York: Crane, Russak, 1971– .

Sayer, Karen and John Dowling. *1984 National Directory of Audiovisual Resources on Nuclear War and the Arms Race*. Ann Arbor, Mich.: Univ. of Michigan, 1984.

Sellers, Robert C., et al. *Reference Handbook of Armed Forces of the World*. New York: Praeger, 1966– .

Stockholm International Peace Research Institute. *World Armaments and Disarmament SIPRI Yearbook 1968–* . Cambridge, Mass.: MIT Pr., 1968– (place and publisher vary).

Swaveley, Peter, ed. *Defense Foreign Affairs Handbook*. Washington, D.C.: Defense and Foreign Affairs, 1976.

The Statesman's Yearbook, 1864– . London: Macmillan, 1864–   (place and publisher vary).

Research Centers and Institutes

Research centers in international relations and institutes are another rich source of information on the topic of military relations and foreign policy. In particular, a number of organizations focusing on strategic studies have emerged in the past few decades. Through journals, conferences, papers, reports, contract work, and so forth, these organizations have strongly affected the literature on the military in general and nuclear studies in particular. While these groups vary in size and importance, they represent a continuing influence over academic thought and public policy. Listed below are some of the more well-known organizations in this area. All of these groups publish annual reports or lists of publications. For persons wanting to order publications or gain information on similar groups, two reference books are essential: Joseph C. Kiger, ed. *Research Institutions and Learned Societies* (Westport, Conn.: Greenwood, 1982), and Mary Michelle Watkins and James A. Ruffner, eds., *Research Centers Directory,* 9th ed. (Detroit: Gale, 1984). Additionally, the PAIS and SSCI indexes should be used to gain access to reports having corporate authors.

Air University Center for Aerospace Doctrine, Research and Education. Maxwell Air Force Base, Ala.
    Established in 1983, the center consists of three organizations: Airpower Research Institute, Air Force Wargaming Center, and Air University Press. The center is under military control; the Center's research is published in reports, monographs, and books.
American Enterprise Institute for Public Policy Research (AEI). Washington, D.C.
    Established in 1943 to provide scholarly economic information, it has grown to a multipurpose research institute. Research is done in ten subject areas, including foreign policy and defense policy. The Institute publishes a series of works including the journal *AEI Foreign Policy and Defense Review,* reprints, and newsletters. Considered to be an important conservative think tank.
Brookings Institution. Washington, D.C.
    Incorporated in 1927 to produce research on broad social, political, and economic issues. Research activities are conducted in economics, government, and foreign policy sections. It publishes annual analyses of the federal budget (a publication entitled *Setting National Priorities,* as well as periodic reviews of national security concerns including nuclear strategy and nonproliferation. It is considered an important influence on public policy.
Carnegie Endowment for International Peace. New York.
    Established in 1910 to "hasten the abolition of international war." Over the years, it has distributed books to libraries, established international law

journals, and sponsored studies and educational conferences on the role of the military in modern society, law, and diplomacy.

Center for Defense Information. Washington, D.C.

Independent educational center focusing on a wide variety of defense issues, including the arms race, military balance, military spending, arms control, arms transfers, and so on. It publishes reports, books, monographs, and the periodical *The Defense Monitor.*

Center for Naval Analyses. Alexandria, Va.

The CNA is a separately incorporated group operating under a contract between the Hudson Institute and the U.S. Navy. The Center conducts research for the Navy and Marine Corps, including such areas as weapons research and development, military operations, force structures, and future strategies. Publishes *CNA Technical Reports* and *Naval Abstracts.*

Council on Foreign Relations. New York.

Incorporated in 1921 to emphasize political aspects of foreign affairs. It has been home to politicians, policymakers, business representatives, and academicians. Publishes the prestigious journal *Foreign Affairs* and the *Foreign Affairs 50 Year Bibliography* (1973), as well as other titles in the areas of international relations.

Hoover Institution on War, Revolution, and Peace. Palo Alto, Calif.

This is the largest private repository of twentieth-century political documents, including the papers of diplomats, politicians, military leaders, and international organizations dealing with war and peace. It publishes over thirty scholarly works a year. Issues an annual report.

Hudson Institute. Croton-on-the-Hudson, New York.

Established in 1965 by Herman Kahn and Max Singer. It has had a major impact, primarily through the work of Kahn and his associates, on nuclear war planning (such as escalation scenarios). It does contract and independent scholarly research on such topics as civil defense, missile defense programs, and arms control.

Institute for Policy Studies. Washington, D.C.

This is an independent research and educational institute. Its focus is on national security and foreign policy, American-Soviet arms race, as well as a wide range of international issues (for example, the economics of development, racism and sexism, and international inequality). Considered an important liberal to radical think tank.

Rand Corporation. Santa Barbara, Calif.

Established in 1946 as a private advisory corporation to the military to "study and [do] research on the broad subject of intercontinental warfare other than surface with the object of recommending to the Army Air Force preferred techniques and instrumentalities for this purpose." This organization has been the dominant private organization involved in national security affairs over the past three decades. Pioneered operations research and heavily influenced systems analysis in military strategy. Many of the important nuclear analysts and strategists (such as Wohlstetter and Enthoven) worked at Rand. While it is less dependent today on military research, it remains an influential force in discussions of current and future military relations and foreign policy.

Peace Studies

A somewhat contending perspective to strategic studies has been termed peace and world order studies. A mixture of groups and organizations, generally supportive of disarmament and dedicated to finding alternative models of conflict resolution and political structures, has emerged in the past few decades. Many publish newsletters, pamphlets, and other types of periodicals. Two excellent guides to these groups are Barbara J. Wein, ed., *Peace and World Order Studies*, 4th ed. (New York: World Policy Institute, 1984) and Robert Woito, *To End War: A New Approach to International Conflict* (New York: Pilgrim, 1982). The Wein book is primarily a curriculum guide to a wide range of global issues (such as hunger, human rights, and militarism); a resource section provides annotated information on funding sources, films, periodicals, organizations, and bibliography. The Woito book discusses ideas, contexts, and actions consistent with ending war. Over 2,000 sources are annotated and generally categorized by subject and whether supportive of ending war. The "action" section provides an annotated listing of world affairs organizations classified by type and purpose. A list of periodicals focusing on world affairs is categorized by topic or emphasis.

United Nations Publications

The United Nations publishes a large number of reports, proceedings, documents, and so on. These publications are covered in Mary Birchfield, comp. and ed., *The Complete Guide to United Nations Sales Publications 1946-1978*, 2 vols. (Pleasantville, N.Y.: UNIFO Publishers, 1982), and *UNDOC: Current Index. United Nations Documents*, a monthly index with yearly cumulations. The index provides a checklist of documents and publications with subject, author, and title indexes. Three publications of the Disarmament Commission are of particular note: (1) the *United Nations Disarmament Yearbook* (1977–); (2) *Disarmament: A Periodic Review by the United Nations* (1978–); and a cumulation series, *Reports and Studies: The United Nations and Disarmament, 1940-1970 and 1970-1975* (1970 and 1976). A wide series of reports, documents, and so on on the two special Disarmament Conferences and the topics of nuclear weapons and nuclear proliferation are available. Specific reference to these materials can be found in the UNDOC index.

Glossary

ABM system: *see* anti-ballistic missile system.

ACDA: *see* Arms Control and Disarmament Agency.

anti-ballistic missile system: a system, typically ground-based, that includes radars and nuclear and nonnuclear missiles theoretically capable of detecting, tracking, and destroying incoming offensive missiles. Agreement limiting development of ABM systems was made between United States and U.S.S.R. as part of SALT I. *See also* ballistic missile defense.

anti-satellite systems: technologies oriented toward destroying satellites used in intelligence-gathering or possibly in actual fighting. ASAT and the ballistic missile defense system (q.v.) are the two major initiatives in space weapons. *See also* Strategic Defense Initiative.

Arms Control and Disarmament Agency: an agency of the federal government whose primary mission involves information dissemination and negotiations regarding nuclear weapons.

ASAT: *see* anti-satellite systems.

ballistic missile: a missile characterized by a free-falling trajectory heavily influenced by gravity. Classified on the basis of range (short-range, intermediate, and intercontinental), it may be land- or sea-launched. *See also* intercontinental ballistic missile, submarine-launched ballistic missile, and multiple independently targetable re-entry vehicle.

ballistic missile defense: a theoretical space-based defense system, although it can include ground-based anti-ballistic missile systems. The term is often used interchangeably with anti-ballistic missile defense system (q.v.). More recently, BMD has been viewed as the central focus of the U.S. Strategic Defense Initiative (q.v.). BMD and anti-satellite systems (q.v.) are the two major initiatives in space weapons.

BMD: *see* ballistic missile defense.

C3I (pronounced "C cubed I"): command, control, communication, and intelligence. The term generally refers to systems of detection and response to political and military conflict, including management of military forces during peace and war.

CEP: *see* circular error probability.

251

circular error probability: a measure of accuracy for a given missile system: if the radius of a circle is centered on a target, one half of the weapons aimed at the target will be expected to fall within the radius. A CEP of 500 yards means that one half of the warheads are expected to land within a radius of 500 yards of the target.

counterforce weapons: nuclear weapons directed toward an opponent's military personnel and installations. A counterforce strategy involves targeting the opponent's military forces either to initiate a nuclear attack or to retaliate. For contrast, *see* countervalue weapons.

countervalue weapons: nuclear weapons targeted toward an opponent's civilian population or industry. A countervalue strategy involves targeting the opponent's civilian population or industry either to initiate a nuclear attack or to retaliate. For contrast, *see* counterforce weapons.

cruise missile: a pilotless missile, armed with conventional or nuclear warheads and launched from an airplane, submarine, or land-based platform. Because it is highly mobile and can be delivered at low altitudes, the cruise missile is an important offensive system that is difficult to defend against.

deterrence: a theory of conflict management in which an opponent is "dissuaded" from initiating an attack by the threat of unacceptable retaliation. Such a theory has been associated with a variety of military and political nuclear policies. *See* mutual assured destruction.

electromagnetic pulse: a wave resulting from a nuclear explosion, particularly one detonated at high altitude, that can severely disrupt and damage electrical equipment. EMP is seen as a threat to command, control, communication, and intelligence systems, significantly reducing the effectiveness of a nation to respond to a nuclear threat.

EMP: *see* electromagnetic pulse.

EMT: *see* equivalent megatonnage.

equivalent megatonnage: a measure of the effectiveness of a nuclear explosion. It is calculated by taking the yield (q.v.) to the two-thirds power. EMT is the preferred measure of destruction when the amount of area destroyed is at issue (e.g., a city). Other measures are used when the concern is with the ability to destroy hardened targets (e.g., missile silos).

Federal Emergency Management Administration: a federal agency whose mission is to coordinate plans for natural and military disasters. Largely associated with civil defense planning, FEMA incorporates the former Defense Civil Preparedness Agency (DCPA), which originally administered civil defense.

FEMA: *see* Federal Emergency Management Administration.

first strike: an initial attack with nuclear weapons. The term sometimes is also associated with a counterforce strategy (*see* counterforce weapons). The prospect that an initial attack could prevent effective retaliation is termed a "debilitating first strike." For contrast, *see* second strike.

fission: a process in which the nucleus of a heavy atom splits into lighter nuclei, thereby releasing vast amounts of energy. Atomic bombs rely on the fission of uranium or plutonium. Also, all fusion (hydrogen or thermonuclear) weapons use a fission weapon to trigger them. (*See* fusion.) Fission reactions release much more energy than conventional explosions.

fusion: a process in which light atoms are combined to form a heavier atom and in the process release vast amounts of energy. Hydrogen or thermonuclear bombs rely on a fusion reaction initiated by a fission weapon. (*See* fission.) Fusion reactions release much more energy than fission reactions.

horizontal proliferation: the spread of nuclear weapons to non-nuclear nations. *See also* vertical proliferation.

IAEA: *see* International Atomic Energy Agency.

ICBM: *see* intercontinental ballistic missile.

intercontinental ballistic missile: a ground-based missile having a range of at least 5,500 kilometers; each ICBM can carry several warheads. *See also* ballistic missile; multiple independently targetable re-entry vehicle; submarine-launched ballistic missile; Triad.

International Atomic Energy Agency: a multinational organization whose primary mission is to promote nuclear energy. It also establishes safe-guards to ensure that nuclear energy will not be used for military purposes.

International Peace Research Association: a scientific research organization associated with Ohio State University that studies problems of war and promotes peace education.

IPRA: *see* International Peace Research Association.

kiloton: a measure of the yield of a nuclear weapon. It is equivalent to 1,000 tons of TNT. Many early nuclear weapons, including those used in Hiroshima and Nagasaki, were of this size. Today, weapons of this yield are primarily characterized as tactical or battlefield weapons. For contrast, *see* megaton. *See also* equivalent megatonnage; tactical weapons.

MAD: *see* mutual assured destruction.

megaton: a measure of the yield of nuclear weapons. It is equivalent to one million tons of TNT, or a thousand times the yield of a kiloton (q.v.). A megaton is typically the unit for measuring the yield of strategic, as opposed to tactical, nuclear weapons. *See also* equivalent megatonnage; strategic weapons.

MIRV: *see* multiple independently targetable re-entry vehicle.

missile experimental (MX): the new generation of American intercontinental ballistic missiles, which include both single and multiple warhead missiles.

multiple independently targetable re-entry vehicle (MIRV): a system in which two or more nuclear warheads carried by a single ballistic missile are capable of being delivered to different targets. *See also* ballistic missile; intercontinental ballistic missile; submarine-launched ballistic missile.

mutual assured destruction (MAD): a theorized capacity of both the United States and the U.S.S.R. to inflict ''unacceptable'' damage on the other after suffering a nuclear attack. *See also* second strike.

MX: *see* missile experimental.

National Security Council: an intragovernmental agency established after World War II to advise the president of the United States on a wide range of geo-political and military matters.

NATO: *see* North Atlantic Treaty Organization.

no first use: a pledge or agreement not to be the first nation to initiate use of nuclear weapons in a conflict. The Soviet Union currently advocates such a pledge while the United States and NATO reject it.

to non-nuclear states as well as negotiations among nuclear states to control and reduce nuclear arsenals.

North Atlantic Treaty Organization: a military alliance formed after World War II whose primary function is the defense of Western Europe.

NPT: *see* Non-Proliferation Treaty.

NSC: *see* National Security Council.

SAC: *see* Strategic Air Command.

SALT: *see* Strategic Arms Limitations Talks.

SDI: *see* Strategic Defense Initiative.

second strike: a retaliatory (usually nuclear) attack following an opponent's first strike. Theoretically, a second strike capability would deter an opponent from initiating an attack as the aggressor would still face an unacceptable response. *See also* mutual assured destruction. For contrast, *see* first strike.

Single Integrated Operations Plan: an evolving plan for control and use of U.S. nuclear forces. Established during the Eisenhower administration, the plan as developed by the Reagan administration is called SIOP-6.

SIOP: *see* Single Integrated Operations Plan.

SIPRI: *see* Stockholm International Peace Research Institute.

SLBM: *see* submarine-launched ballistic missile.

START: *see* Strategic Arms Reduction Talks.

Stockholm International Peace Research Institute: an organization established in 1966 to research problems associated with war and peace.

Strategic Air Command: initially, the primary military organization for the delivery of a nuclear weapon by bomber. Currently it is one of three delivery systems. *See* Triad.

Strategic Arms Limitations Talks: negotiations begun in 1969 by the United States and the U.S.S.R. to control nuclear forces of both countries. Agreements, understandings, and treaties associated with SALT have resulted, not all of which have been ratified by the U.S. Senate. *See also* strategic weapons.

Strategic Arms Reduction Talks: negotiations begun in 1982 by the United States and the U.S.S.R. on the strategic arsenals of each country. START succeeds the Strategic Arms Limitations Talks (q.v.).

Strategic Defense Initiative: the Reagan administration's proposed development of a space- and land-based system of defense against strategic nuclear weapons. While SDI is a policy objective, it remains a theoretical possibility rather than an existing technological reality. *See also* anti-ballistic missile system; anti-satellite systems; ballistic missile defense.

strategic weapons: weapons and/or forces capable of affecting an opponent's warring ability. For example, a strategic nuclear weapon is often viewed as having the capacity of hitting a military or civilian target in the opponent's homeland. Approximately 20,000 weapons in the nuclear arsenals of the United States and the U.S.S.R. are categorized as strategic. For contrast, *see* tactical weapons.

submarine-launched ballistic missile: the sea-based leg of the three primary means of delivering nuclear weapons. *See also* ballistic missile; intercontinental ballistic missile; multiple independently targetable re-entry vehicle; Triad.

means of delivering nuclear weapons. *See also* ballistic missile; intercontinental ballistic missile; multiple independently targetable re-entry vehicle; Triad.

tactical weapons: weapons and/or forces for combat with an opponent. Tactical nuclear weapons are normally characterized as having a limited range and yield, in contrast to the larger range and yields of strategic weapons (q.v.).

thermonuclear weapon: a nuclear weapon in which thermonuclear fusion reactions create the explosive energy released. Hydrogen bombs are thermonuclear weapons. *See* fission; fusion.

Triad: the tripartite division of nuclear weapon delivery systems, including land-based intercontinental ballistic missiles (q.v.), submarine-launched ballistic missiles (q.v.), and bombers.

vertical proliferation: the increase in the numbers of nuclear weapons systems by nuclear nations. *See also* horizontal proliferation.

Warsaw Treaty Organization: a military alliance formed after World War II whose primary function is the defense of Eastern Europe and the Soviet Union.

WTO: *see* Warsaw Treaty Organization.

yield: the force of a nuclear explosion. It is generally expressed as the equivalent of energy produced by tons of TNT. *See also* equivalent megatonnage; kiloton; megaton.

Author-Title Index

Harf, James F. National Security Affairs, 215.
Harkavy, Robert. The Arms Trade and International Systems, 152; Arms Transfers in the Modern World, 153.
Harlan, Cleveland. "A Strategy for the United States," 144.
Harvard Nuclear Study Group. Living with Nuclear Weapons, 214.
Harvey, Mose L. The Role of Nuclear Forces in Current Soviet Strategy, 189.
Harwell, Mark A. Nuclear Winter, 110.
Has Man a Future? Russell, 197.
Head, Richard G. American Defense Policy, 50.
Hearings on the Strategic Defense Initiative. U.S. Congress, 215.
"Heidegger and Nuclear Weapons." Smithka, 198.
Heilbrunn, Otto. Conventional War in the Nuclear Age, 216.
Heim, Michael. "Reason as Response to Nuclear Terror," 196.
Hekir, J. Bryan. The New Nuclear Debate, 195.
Henkin, Louis. Arms Control, 185.
Herman, Theodore. Peace and War, 216.
Hermann, Charles F. International Crises, 189.
Hersey, John. Hiroshima, 33.
Herzog, Arthur. The War-Peace Establishment, 193.
Hewlett, Richard G. Atomic Shield, 1947–1952, 32; History of the U.S. Atomic Energy Commission, 32; The New World, 1939–1946, 32.
Heyer, Robert. Nuclear Disarmament, 216.
High Frontier. Graham, 35, 214.
Hildreth, Steven A. Modern Weapons and Third World Powers, 152.
Hill, Kim Quaile. "Domestic Politics, International Linkages, and Military Expenditures," 145.
Hill, Walter W. "A Time-Lagged Richardson Arms Race Model," 145.
Hillenbrand, Martin J. "NATO and Western Security in an Era of Transition," 189.
Hines, Neal O. Proving Ground, 106.
Hiroshima. Fogelman, 33.
Hiroshima. Hersey, 33.
Hiroshima and Nagasaki. Committee for the Compilation of Materials, 32, 104.
Hiroshima/Nagasaki. 33.
History of Logic. Dumitriv, 81.
A History of Militarism. Vagts, 19.
History of the U.S. Atomic Energy Commission. Hewlett, 32.

Hitch, Charles J. The Economics of Defense in the Nuclear Age, 152.
Hoag, Malcolm W. "NATO," 185.
Hobson, J. A. Imperialism, 18.
Hoenig, Milton. U.S. Nuclear Forces and Capabilities, 36, 141.
Holden, Constance. "Military Grapples with the Chaos Factor," 111.
Hollenbach, David. Nuclear Ethics, 193.
Holloway, David. The Soviet Union and the Arms Race, 145.
Holst, Johan J. Beyond Nuclear Deterrence, 145, 189, 216; Why ABM? 37.
Holsti, Ole R. Crisis, Escalation, War, 190.
Holton, Gerald. "Arms Control," 184.
Hook, Sidney. The Fail-Safe Fallacy, 196.
Hopkins, Terrence. Processes of the World System, 18.
Horsburgh, H. J. N. Non-Violence and Aggression, 218.
Hostage America. Cutright and Dentler, 185.
"How Much Can 'The Just War' Justify?" Wells, 198.
How Much Is Enough? Enthoven and Smith, 53, 145, 189.
How Much War in History. Beer, 17.
How to Think about Arms Control and Disarmament. Dougherty, 145.
Howard, Michael. Theory and Practice of War, 19.
Howe, Russell W. Weapons, 152.
Huisken, Ronald. Arms Uncontrolled, 139.
The Human Imperative. Alland, 17.
The Human Zoo. Morris, 18.
"Humanism, Ontology, and the Nuclear Arms Race." Zimmerman, 198.
Hunter, John E. "Mathematical Models of a Three-Nation Arms Race," 145.

The Idea of Disarmament! Geyer, 195.
Ikle, Fred. "Can Nuclear Deterrence Last over the Century?" 190.
The Illogic of American Nuclear Strategy. Jervis, 214.
"Illusion of Survival." Gieger, 90–91.
"The Impact of Militarization on Development and Human Rights." International Peace Research Association, 152.
Impact of New Technologies on the Arms Race. Feld, 145.
Imperialism. Hobson, 18.
Imperialism. Lenin, 18.
Imperialism. Magdoff, 18.
"The Implications of the Atomic Bomb for International Relations." Viner, 183.

Subject Index

William C. Gay is associate professor of philosophy at the University of North Carolina at Charlotte, where he teaches courses in contemporary philosophy, war and peace studies, and critical thinking. Gay is the editor of a double issue of *Philosophy and Social Criticism* entitled "Philosophy and the Debate on Nuclear Weapon Systems and Policies." He also has chapters in *Nuclear War: Philosophical Perspectives* (Peter Lang, 1985) and the forthcoming *Religion and Philosophy in the United States* (Die Blaue Eule, 1987), and, with R. E. Santoni, a chapter in the forthcoming *Critical Bibliography of Genocide* (Mansell, 1987).

Michael A. Pearson is assistant professor of sociology at the University of North Carolina, Charlotte, where he teaches couses in social movements, American minority groups, and war and peace studies. He has published articles and chapters in *Social Science Quarterly*, *Sociological Symposium*, *The American Military* (Aldine, 1970), and *Essays in the Sociology of Social Control* (Gowes, 1986).

The editor of the series, John H. Whaley, Jr., has a master's degree in library science and a Ph.D. in history. He has been a librarian since 1974 and has had ten years' experience in reference. At present he is Associate Director for Collection Management at Virginia Commonwealth University in Richmond.